CREATIVE INVESTIGATIONS
IN EARLY SCIENCE

Angela Eckhoff, PhD

www.gryphonhouse.com

Copyright ©2018 Angela Eckhoff

Published by Gryphon House, Inc.
P. O. Box 10, Lewisville, NC 27023
800.638.0928; 877.638.7576 (fax)
Visit us on the web at www.gryphonhouse.com.

All rights reserved. No part of this publication may be reproduced or transmitted in any form or by any means, electronic or technical, including photocopy, recording, or any information storage or retrieval system, without prior written permission of the publisher. Printed in the United States. Every effort has been made to locate copyright and permission information.

Cover images used under license of Shutterstock.com and courtesy of the author. Interior images courtesy of the author and Shutterstock.com.

Library of Congress Cataloging-in-Publication Data
The cataloging-in-publication data is registered with the Library of Congress for
ISBN 978-0-87659-804-7.

Bulk Purchase

Gryphon House books are available for special premiums and sales promotions as well as for fund-raising use. Special editions or book excerpts also can be created to specifications. For details, call 800.638.0928.

Disclaimer

Gryphon House, Inc., cannot be held responsible for damage, mishap, or injury incurred during the use of or because of activities in this book. Appropriate and reasonable caution and adult supervision of children involved in activities and corresponding to the age and capability of each child involved are recommended at all times. Do not leave children unattended at any time. Observe safety and caution at all times.

Contents

Dedication......................................iv

Introduction.....................................1

Chapter 1: Physical Science:
    Understanding Matter and Physical Properties..............17

Chapter 2: Physical Science: Physical and Chemical Changes.....34

Chapter 3: Life Science: Growth and Change..................54

Chapter 4: Earth Science: Conservation and Sustainability.......71

Chapter 5: Earth Science: Earth and Space Systems............86

Index..98

Dedication

A special thank you to the early childhood science methods students I've worked alongside during my career. The classes we've shared exploring how to encourage children to experiment, question, get messy, and play provided me with the inspiration to make this book a reality. Thank you!

Introduction

Inquiry-Based Learning in Early Science

Tommy and three of his classmates are sitting together in the art area working on paintings of imagined, mythical creatures known as zoomorphic animals. Zoomorphic creatures are fantastic creations consisting of various parts of real animals put together in novel ways. This class of five- and six-year-olds has been studying the characteristics and habitats of animals for several weeks as part of the science curriculum. Tommy's teacher, Mr. Brown, had recently shared a story that features zoomorphic animals, *If I Had a Gryphon* by Vikki VanSickle, and many of the children expressed an interest in developing their own mythical creatures. To support this exploration, Mr. Brown stocked the classroom's science center with pictures of various animals for the children to use to explore, as well as with paints, paper, colored pencils, and a variety of markers to use in the drawing and painting phases of the project. He encouraged the students to begin by sketching draft drawings and working to modify those drawings over the course of several days. Once they were satisfied with their sketches, the children were then encouraged to use the arts media to add color and definition to the images.

The subject of Tommy's zoomorphic work is a combination of several of his favorite real-life animals and features the head of an elephant, the wings of an eagle, and the sharp spines of an iguana. Tommy's zoomorphic creature is a fantastic combination of animal characteristics that give his creature "super powers to be the strongest and fastest"—to be as strong as an elephant, to fly like an eagle, and to have the natural defenses of an iguana. As Mr. Brown comes over to talk about the boys' work, Tommy excitedly draws Mr. Brown's attention to his distinctive creature by pointing out its component parts. Mr. Brown and Tommy discuss the parts of the drawing that are associated with each animal that was used as inspiration. "Okay, Tommy," says Mr. Brown, "I have a challenge for you. If your creature has an elephant head but the body of an eagle, what does it eat?" Tommy replies excitedly that his elephant head likes to eat peanuts, not fish or squirrels like eagles do. Mr. Brown laughs and asks Tommy, "Where will this creature sleep? Will it have a nest like an eagle?" Tommy quickly answers yes and says, "He is little like an eagle so he needs to sleep up high in a nest away from big animals."

Tommy's zoomorphic creature experience provided him with opportunities to bring together his understanding of the needs and characteristics of animals as well as his creative-thinking and visual-art skills. Mr. Brown has created a learning environment that encourages and supports connections between science content and the children's desire to engage in creative experiences. The exploration of zoomorphic creatures provides unique opportunities for each child to develop his own interpretation and encourages the children to reflect and draw upon their knowledge of the physical characteristics of animals.

Mr. Brown enhanced this experience for his students by encouraging engagement with one another through the sketching and research experiences with the animal photographs. He also extended the children's individual work by asking prompting questions.

Young Children Are Scientists

All children are scientists; during the early childhood years, children naturally engage in the scientific processes of observation, manipulation, experimentation, and exploration. The natural curiosity of young students provides early childhood educators with an entry point from which to build classroom science experiences. Inquiry-based early childhood science education capitalizes on the interests young children demonstrate as they explore the world around them. Through creative, hands- and minds-on experiences, early childhood educators can encourage children to construct understanding of earth, life, and physical sciences in ways that are personally meaningful. The science experiences described in this book are based on the 5E inquiry model of instruction (engage, explore, explain, extend/elaborate, and evaluate) as well as the content recommendations from the *Next Generation Science Standards* (NGSS).

This book is designed to provide early childhood educators with pedagogical practices, science content knowledge, and lesson ideas that scaffold young children's experiences with earth, life, and physical science, while also building inquiry and creative-thinking skills. This book will broaden your understanding of the relationships among science content, the role of the learning environment, and supportive pedagogical practices in early childhood classrooms. When science experiences build on student interests and understanding and connect to other areas of content learning—literacy, technology, engineering, the arts, mathematics, and social studies—young children are able to experience meaningful, relevant connections among different content areas. This book stresses the importance of encouraging *minds-on* learning experiences in the early childhood classroom through guided and independent investigations, where every child is actively involved in meaningful ways. Early childhood educators have important roles in science-focused experiences and will act as both guide and facilitator throughout the planning, implementation, and assessment of the creative, inquiry-based experiences presented throughout this book. For young children, science experiences involve using tools and a variety of materials, being creative and inventive, developing questions based on observations, exploring problems, and sharing their understanding with others.

Creative Investigations in Early Science will support your development of creative early science experiences in the classroom by helping you:

- understand the links among science content, inquiry-based learning, and project-based learning;

- plan cooperative science lessons that will engage all children in your classroom as individuals or when working in small or whole groups;

- implement classroom experiences that support children's engagement with science content on a daily basis;

- recognize children's understanding, beliefs, and misconceptions of science concepts and utilize that information to support the growth of conceptual knowledge; and

- document children's knowledge development with authentic work samples and classroom artifacts.

Playful Learning

Play is an essential element in explorations of science in early childhood. Through play, young children learn about themselves, their environment, other people, and the world around them. Playful learning encourages children to explore and experiment in situations where they feel comfortable taking risks and delving into the unknown. Children's play in the early childhood classroom can take on many different forms and functions. When children explore, experiment, and cooperate through play, they learn about how the world works. Children need teachers who are supportive of their play and who work to carefully identify play situations where teacher guidance or involvement are welcome and needed.

Young children use their knowledge and understanding by bringing these ideas into their play to further experiment and clarify their understanding. This process is child driven; the role of the adult is one of supporter, guide, and facilitator. The adult meets each child at his own stage of understanding with intentional pedagogical practices that promote questioning and exploration. Teachers can create early childhood classrooms that honor the ways in which children learn and explore by ensuring that young children have ample opportunities for playful learning and exploration. In the role of supporter, guide, and facilitator, the teacher carefully observes children's play and helps encourage children's thinking through questioning and providing additional, supportive materials and opportunities for guided learning.

Guided Inquiry in Early Science Experiences

Inquiry-based learning can play a central role in the development of meaningful learning opportunities as children explore emerging skills in early science. Contrary to traditional notions of the teacher's role as a teller of information, teachers in inquiry classrooms perform the roles of guide, facilitator, and provocateur by asking questions and designing meaningful lessons built on student interests. A teacher's ability to listen to her students is a foundational component of the use of guided inquiry in science explorations. By carefully listening to students and reflecting on their ideas and interests, you will be able to plan and implement engaging and meaningful science explorations with your students that encourage individual expression.

Inquiry-based science experiences in early childhood classrooms are based on the 5E instructional model, where students are first engaged in the topic and then explore using materials and media, which is then followed by opportunities for explanation and elaboration. In the 5E model, both teachers and students work to evaluate ideas and understanding throughout the entire experience (Bybee et al., 2006). Inquiry-based science learning requires planning and intensive engagement on the part of the teacher as well as attentiveness and active engagement on the part of the children. It is recommended that early science experiences incorporate opportunities for exploring both content and inquiry skills. Inquiry-based learning requires these process skills: observation, exploration, questioning, making predictions, using simple tools and technologies, and conducting simple science investigations. In the introduction to each chapter of the book, you will find suggested ideas and practices for each phase of the 5E model based on the content covered in that chapter.

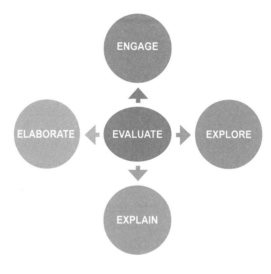

The 5E Model

Engage

Students come to learning situations with prior knowledge, incomplete understanding, and even misconceptions. The *engagement* phase of the model provides opportunities for teachers to find out what students already know or believe about the concept under exploration. This phase also gives the children an opportunity to think about and discuss their thoughts about the concept. The engagement phase is important because it offers an opportunity to capture student interest and inspires young children to want to find out more.

Explore

The *exploration* phase of the inquiry cycle involves the hands-on, minds-on engagement of students. During the exploration phase students will be actively working as individuals or as part of a group investigating materials, ideas, and questions. Time and space are important elements to exploration, so in this phase you will need to provide children with ample time and physical space to conduct their investigations. The exploration phase can take place over several days or even weeks, depending on the concept under investigation and the children's interest.

Explain

The *explanation* phase of the inquiry model provides opportunities for students to connect their prior understanding with their current experiences. Through both verbal (discussion) and physical explanations (drawings, journals, models), the explanation phase helps students develop their conceptual understanding of the science content. Because of the emphasis on sharing understanding, this phase also provides opportunities for you to introduce science language and terms to help support students' explanations. Prompting questions you can pose during this phase may include the following: What did you notice? How can you show us what you know or experienced? Can you tell us more about why you think that happened?

Elaborate

The *elaboration* phase of the inquiry model provides opportunities for children to apply or extend previously introduced concepts and experiences to new situations. In early childhood classrooms, opportunities for elaboration can begin through follow-up experiences in the science center or during paired or small-group experiences. In the elaboration phase, it is important for students to have opportunities to discuss and compare their ideas with others.

Evaluate

In early childhood science experiences, informal observations and interactions with students throughout all phases of the inquiry model are the most appropriate ways to gather information on student understanding. The *evaluation* phase should always be directly connected to students' in-process work rather than the end product of an experience. You can engage your students in the evaluation phase by encouraging them to share their experiences and understanding with others and to listen and respond to the ideas of their peers.

Moving beyond Misconceptions in Science

Misconceptions are ideas that a person may have that are not aligned with accepted scientific views. We all have science misconceptions; many are formed about concepts that are frequently misunderstood. A common misconception you have probably heard more than once is that humans only use 10 percent of their brains. Cognitive scientists have worked for years to change this misconception by stating that there is no scientific evidence to suggest that we use only 10 percent of our brains. In fact, brain imagining demonstrates that when we move, speak, or think about a particular object, brain activity is widely noted in many regions. However, once many of us hear this myth, it stays with us and we may repeat it to others. Like adults, young children frequently have misconceptions. A few common misconceptions they may hold are the following:

- Rain comes from holes in clouds.

- It rains because we need or want it.

- Leaves pick the color they want to change to in the fall.

- Humans are not animals. The moon can only be seen during the night.

- There are four separate moons.

If we spend some time thinking about these misconceptions, we can begin to understand how children develop these ideas. When children look at the sky on a rainy day, they may see clouds that appear to have holes, which would let rain leak out. They also hear adults say things such as, "The tree makes such beautiful colors in the fall." We point out the moon at night to children and read books that connect the moon and nighttime, but we may not take advantage of a clear sky during the day when we could point out the moon. The misconception that there are four different moons could be developed as children view the moon intermittently, only viewing the moon on occasion and noting that it looks different from the time before. Without intending to, we help to create misconceptions in the classroom when we do activities that might serve to reinforce their incomplete understanding. For example, a common phase-of-the-moon lesson involves children using cookies or paper cutouts with small portions removed to represent each phase. As this activity requires students to create and display linearly four separate moons, we can understand how they arrive at a misconception.

An activity that would better reinforce the idea of one moon moving through phases would be one in which the four phase cutouts could be laid on top of each other to show a single moon. Children can visualize that we can only see a part of the moon in each phase and can learn how the full moon is blocked, preventing us from seeing it in its entirety. This activity will also allow the children to see how each phase relates to the previous phase. You may want to follow up this experience by inviting children and families to complete a month-long moon journal in the evenings, which will further deepen the children's understanding. If children have the opportunity to view the moon every evening (or when possible) and to draw what they see for a month, they will begin to understand how the moon goes through the phases gradually. Providing young children with experiences that intentionally build on each other will help avoid instances of reinforcing or developing misconceptions.

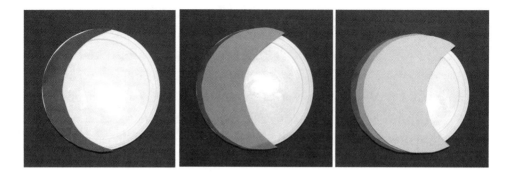

We can work to identify misconceptions by taking the time to talk and listen to students' thoughts and conversations during all phases of the inquiry cycle. In particular, asking children what they already know or understand about concepts during the exploration phase of the cycle will help you to decide how best to proceed in your instruction. It's challenging to change misconceptions because they are based on our held beliefs, and we don't readily shed beliefs without prompting. Fortunately, we can work with young children and provide them with many opportunities to challenge their thinking. Having lessons in which students are interested will help support their willingness to think deeply.

The concepts that you are teaching need to be appropriate and at a level that children can understand. For example, the concept of seasonal change is very complex. When we break it down into conceptual knowledge that is appropriate for young learners, we can focus on experiences where children observe and describe the characteristics of living things, compare the growth of a person to the growth of a plant and an animal, and describe the basic needs and the basic life processes of each. This foundational knowledge will help them understand in later grades why we experience seasonal change. Building conceptual knowledge in science is a long-term endeavor, and the early childhood years are the time where we can help develop children's initial understanding, skills, and dispositions.

Building Creative Science Experiences in the Classroom

Early childhood educators have essential roles in the development of children's creative-thinking skills because they can create either supportive classroom environments or classrooms in which children's creative skills are stifled. To incorporate creative learning experiences in the classroom, teachers must design lessons that include opportunities for critical thinking and reflection while also maintaining a focus on student interest. In addition, teachers must recognize that creativity is a learning process that encourages social interaction and promotes individual ownership of ideas. In the classroom, creativity is part of the learning process based on children's interests, involves reflection and interaction with other children and adults, and requires children to document and report on their thinking and experiences. When young children are provided opportunities to personally engage with challenging, reflective learning experiences, they are building critical- and creative-thinking skills.

The lesson ideas and classroom vignettes shared throughout this book incorporate opportunities to build children's inquiry process skills and their understanding of earth, life, and physical science while also promoting their creative-thinking skills. Each lesson includes critical elements of inquiry and creative thinking—open-ended tasks, opportunities for social interaction, and opportunities for reflection and elaboration. Open-ended tasks provide young learners with opportunities to experiment with new ideas and engage in inquiry. Because open-ended tasks promote idea experimentation, they encourage children to focus on the processes of learning rather than the need to arrive at a solitary correct answer. Gaining experience with idea experimentation will help support children's acceptance of ambiguity and the willingness to make mistakes, allowing them to gain confidence in their problem-solving abilities. Likewise, providing opportunities for small-group work and social interaction is a crucial component of creative thinking. Working in pairs or small groups will help to promote brainstorming and allow children to learn from and with each other. Such tasks will also support children's experiences with reflection and idea elaboration. These skills are important cognitive tools that allow children to learn from their own experiences and examine their own learning processes.

Recommended Practices and Content Coverage in Early Science Experiences

The content of the lessons presented in each chapter of this book are based on the guiding recommendations in the *Next Generation Science Standards* from National Academies Press. While these standards do not speak directly to young children in the preschool years, you can use these guidelines to help determine the types of experiences you can develop in your classroom. This will give your students a solid foundation in both content understanding and experiences and will help them engage in creative thinking processes and inquiry-based learning. In every chapter you will find a section that introduces you to the relevant core ideas from the *Next Generation Science Standards* related to the content in that chapter. Every lesson presented in this book is designed to encourage you to explore and implement the types of inquiry-based science experiences that will build children's thinking, exploration, questioning, and documentation skills in addition to curricular content knowledge. Every lesson will ask you to carefully consider your interactions with young children as well as the classroom environment. The interplay among children, teachers, and the classroom environment are all central to the process of learning. The concept of *possibility thinking* encourages teachers to consider the effect that asking questions, play, supportive classrooms, imagination, innovation, and risk taking have on the processes of thinking and learning.

Possibility thinking—A dynamic interplay between children and teachers (Craft et al., 2012)

Posing questions—questions from children are acknowledged and celebrated by teachers; teachers' questions encourage inquiry

Play—opportunities for extended play periods

Immersion—immersion in a benign environment free from criticism and mockery

Innovation—teachers closely observe innovations in student thinking to prompt and encourage

Being imaginative—ample opportunities to meld imagination and curriculum content

Self-determination and risk-taking—deep involvement and risk-taking are encouraged by both children and teachers

Promoting Creative, Inquiry-Based Learning in Science

Classroom Components	*Supportive Approaches in the Early Childhood Classroom*
Physical Environment	• Flexible spaces with movable furnishings that provide space for exploration, display, and storage, and spaces that can accommodate and adapt for small and large groups
Role of the Teacher	• Provide opportunities for children to document their thinking through drawing, writing, and verbal means • Encourage students to share their thoughts with a large/small group • Ask questions to promote deep thinking and problem solving • Provide materials that can support student inquiry • Closely monitor student thinking and exploration to scaffold experiences

Classroom Components	Supportive Approaches in the Early Childhood Classroom
Peer-to-Peer Relationships	• Provide opportunities for children to share their problem-solving experiences and encourage and support children's use of inquiry-based and creative thinking • Provide opportunities for children to ask questions, design experiments/plans, work in pairs/small groups, test ideas, and document their experiences
Structure of Technology and Engineering Experiences	• Provide opportunities for children to connect science to other content areas, work on problems and projects for extended periods of time, and revisit previous experiences and lessons multiple times to encourage mastery and promote confidence
Parent and Community Engagement	• Provide opportunities to connect science experiences into the community and the children's daily lives • Engage families throughout the learning process through regular documentation of children's experiences

Creating Engaging Science Centers

In addition to planning and implementing science experiences in the classroom, it is important to create learning spaces where your students are able to further their own explorations. A science learning center is a good place to invite your students to work individually or in small groups; these centers can be permanent or movable, depending on the interests and needs of your students at any given time. Many science concepts are a natural extension to preschoolers' play in outdoor spaces where children may freely explore life- and earth-science content.

Classroom Components	Teacher Actions
Physical Environment	• Include a variety of natural materials, content-focused books, posters or colorful photos, and child-friendly science equipment (magnifying glasses, scales, rulers)
Role of the Teacher	• Develop a supportive environment for playful learning, experimentation, and risk taking • Closely observe children's play and exploration (formative assessment) • Ask thoughtful questions and provide opportunities to expand and clarify children's thinking
Peer-to-Peer Relationships	• Plan opportunities for collaborative experiences • Demonstrate respect for students' work and ideas • Provide opportunities for play and exploration
Structure of Science Lessons and Experiences	• Plan opportunities for individual and group experiences • Keep a flexible schedule for lesson lengths based on children's responses and interests • Develop extended, project-based science experiences for complex content • Plan opportunities for children to make their thinking visible (exploratory, hands-on experiences; science journals; digital photography) • Extend familiar lessons and concepts to deepen and encourage flexibility of student understanding

Organization of the Book

This book is based on broad categories for early science explorations: matter and physical properties, physical and chemical changes, growth and change, conservation and sustainability, and earth and space systems.

Each chapter begins with a section where you can find background information on physical-, life-, or earth-science content and the processes of inquiry related to each content area. Each chapter also features classroom vignettes to help bring the information on content and pedagogical information to life. Woven throughout the book are science lessons for preschoolers that are built on pedagogical practices for creative, inquiry-based thinking. You will also find children's book recommendations related to each chapter's content.

References

Bybee, Rodger W., et al. 2006. *The BCBS 5E Instructional Model: Origins and Effectiveness.* Colorado Springs, CO: Biological Sciences Curriculum Study.

Craft, Anna, Linda McConnon, and Alice Matthews. 2012. "Child-Initiated Play and Professional Creativity: Enabling Four-Year-Olds' Possibility Thinking." *Thinking Skills and Creativity* 7(1): 48–61.

1
Physical Science: Understanding Matter and Physical Properties

Young children explore their world using their five senses to take in information about the physical properties of objects. These explorations serve as a foundation to understanding objects and the behaviors of those objects under various conditions. A young child licking a frozen treat outside on a warm summer day quickly learns that it can melt before she is finished! These direct, informal experiences allow children to make inferences about the reasonable and sometimes unreasonable explanations of consequences.

Matter is anything that has mass and takes up space. Mass is the amount of matter in an object and differs from weight, which is a measurement of the gravitational pull on an object. Young children can readily explore the physical properties of matter, including size, shape, color, texture, hardness, melting point, magnetism, and whether an object sinks or floats. Explorations with the physical properties of matter provide creative opportunities for inquiry-based learning through making observations, developing predictions, and conducting simple experiments.

The Cycle of Inquiry: Matter and Physical Properties

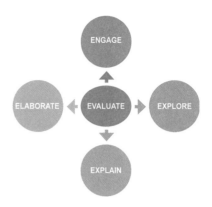

Engage

Questions to Engage: Matter and Physical Properties

Ask these types of questions to engage young learners' interests:

- Which of these objects do you think is heavier? What makes you think that?

- What do you think will happen if we put this ice cube in the sun and put this one in the shade? Why will it melt faster/slower?

- What do you think this object will feel like in your hand?

- Can you describe this object with one word?

Explore

Engaged Exploration: Matter and Physical Properties

To assist children as they work to explore the world around them, provide a space in the classroom that allows them to do the following:

- Sort, compare, and classify objects by their physical properties (color, shape, texture, size, weight, and solid or liquid state)

- Explore water and objects (sink or float)

- Touch objects of varying textures and materials

Explain

Explanation Opportunities: Matter and Physical Properties

Use the following suggestions to help children understand and explore class lessons on a deeper level:

- Display properties charts in the classroom
- Display photographs of experiments with student quotes
- Plan whole-group debriefings that discuss observed changes in matter
- Encourage students to draw, write, and comment in their science journals
- Develop whole-class exploration charts with objects that sink or float, objects that are magnetic or nonmagnetic, and objects that are hard or soft

Elaborate

Science Center Elaborations: Matter and Physical Properties

Engage your students with the following ideas to continue their learning:

- Plan water play indoors or outdoors with a variety of natural and man-made objects to test whether they sink or float
- Test magnets and a variety of materials to learn about magnetism
- Classify a variety of materials to explore the size, shape, color, texture, or hardness properties

Evaluate

Documenting Informal Evaluation: Matter and Physical Properties

Assist your students as they evaluate their learning from the explorations by using the suggested ideas:

- Plan whole-class review sessions
- Have a science night for families and friends
- Display documentation panels of major projects
- Display student portfolios
- Involve students in planning next steps based on their wonderings

Core Ideas in Matter and Physical Properties

Properties of Matter

- Different kinds of matter exist and many of them can be either solid or liquid, depending on temperature.
- Matter can be described and classified by its observable properties.
- Different properties are suited to different purposes.
- A great variety of objects can be built from a small set of pieces.
- Heating or cooling a substance may cause changes that can be observed. Sometimes these changes are reversible, and sometimes they are not.

Source: NGSS Lead States. 2013. *Next Generation Science Standards: For States, by States.* Washington, DC: National Academies Press.

Vignette for Understanding: Melting Color Mixing

Jackson and his friend Aman are having a lively discussion about which colors to use in their partner painting. Their teachers, Mrs. Kelleman and Ms. Donnelly, have invited the children to work in pairs today to create a color-mixing painting. They have frozen various colors of water in large ice cube trays with craft sticks to create a paint pop. Each pair is invited to select two different colors to create their partner painting. Jackson and Aman choose a dark red and a bright yellow paint pop to use in their artwork. Before they begin painting, Ms. Kelleman asks the boys to predict what colors they will see in their painting as well as what will happen to their paint pops while they paint. Aman immediately responds, "They're going to melt; we have to hurry," and Ms. Kelleman asks the pair why they think the ice will melt. Both boys decide that the pops will melt because it's not cold in the classroom. Ms. Kelleman responds, "That's a great idea. We took them out of the freezer, which kept them frozen, but you're right: It's much warmer in the classroom than the freezer. You said you have to hurry; do you think they'll melt fast or slow?" The boys unanimously agree, "Fast!" Ms. Kelleman encourages them to pay

Physical Science: Understanding Matter and Physical Properties

close attention to how their colors mix and how fast their paint pops melt as they work together. The boys get to work pushing their paint pops around the paper, noticing how their colors are blending together on the page.

Reflection

Jackson and Aman had the opportunity to freely explore during this open-ended experience with color and the freezing and melting of water. When Ms. Kelleman designed the lesson, the decision she made to have the students work in pairs prompted the children to talk to each other about their ideas, which was an important part of the boys' experience. The inquiry process in this experience includes opportunities for students to make observations, explore, question, make predictions, and conduct simple science investigations.

Matter and Physical Properties across the Curriculum: Planning Tips

Experiences with matter and physical properties can involve many different types of materials that are already available in your classroom. Having a science center with opportunities to explore is a good pedagogical strategy for encouraging children to build and explore during independent investigations. Ample space and flooring considerations will be necessary when planning a science center for matter explorations because the children will be exploring water and other materials that, while messy, are necessary for this science content. Such a center can involve a water table; a sand table; and materials of various shapes, sizes, colors, and textures. Connections with mathematics can be made as children sort and classify objects based on their physical properties.

Lesson Ideas

Object Grab Bag

Topic:

Describe and sort objects by their physical properties: color, shape, texture, size, and weight.

Objective:

Children will explore the properties of objects and describe a selected object to their peers.

Materials:

Small paper sack (or any opaque bag)

Small items from the classroom that have properties the children can describe (colorful building blocks or math counters; crayons; craft sticks; glass or plastic cabochons of various sizes, colors, and shapes; stones; and tiles of various sizes and shapes). Items should be small enough to be hidden inside a child's clasped hands.

> **Creativity Skills:**
>
> Exploration
>
> Visualization
>
> Communication/ collaboration

Overview:

This game can take place during whole-group time or in small groups.

Activity Steps:

1. Prior to starting the game, gather enough items for each child to have at least one turn selecting an item. Talk with the children about how they are going to take turns selecting a mystery object from the bag that they will describe to the other students without saying what the object is. The class will then take turns guessing the item based on the students' description. If this is their first time playing the game, be sure to play a practice round or two so that the students understand how to describe objects and when to make guesses.

2. Invite a child to begin the game by selecting the first object without looking in the bag. Encourage the child selecting the object to give one descriptive clue at a time. Have the child face you with her back to the class so that she can pull the object out of the bag and peek at it before covering it with her hands. Once she is ready, she can turn to face the class for the guessing to begin.

3. You may want to prompt the child to describe her item until the class gets used to providing descriptive clues. Prompting questions can include the following: Can you describe the color of your object? Can you describe the shape of your object? Is your object heavy or light? How does your object feel when you touch it? Where do we play with that object in the classroom?

4. After the children guess the object, invite the next child up to select an object until each child has had a turn.

Documentation:

Take anecdotal notes about the children's abilities to describe the properties of selected objects.

Extension Lesson:

This lesson can be extended by using specific groups of objects, such as objects from nature, to extend science explorations taking place within the classroom.

Dancing Liquids

Topic:

The properties of objects can be described.

Objective:

Children will explore how various liquids react when they are placed on a variety of surfaces.

Creativity Skills:

Exploration

Communication/ collaboration

Materials:

Wax paper/parchment paper

Aluminum foil

Construction paper or card stock

Cellophane or plastic wrap

Small container of water

Small container of liquid soap

Small container of vegetable oil

Small pipette or medicine dropper

Scissors (adult use)

Overview:

This is an exploratory experience that can be done in a whole or small group and then moved to the science center for students to explore further.

Activity Steps:

1. Precut small squares (3" x 3") of each of the surface materials. You will need to have at least three of each type for the exploration of all three liquids.

2. Invite the children to feel each surface type and use descriptive language to describe the surface (shiny, smooth, sticky, and rough).

3. Show the children each of the liquids. Ask them to think about which liquid is thinnest and which is thickest.

4. Talk to the children about how they will be conducting an experiment to see what happens when a drop of each liquid is placed on the different surfaces. Encourage them to make predictions—

What do you think will happen when we put water on the paper? What will happen when the oil is dropped on the foil?

5. Invite the children to help drop the liquids one by one on each surface. Take time between drops to draw their attention to the reactions of the liquids on the surfaces.

6. Invite them to think about why the liquids had differing reactions on the various surfaces: Why do you think the water was absorbed on the construction paper and stays in a drop on the foil? As they think about the reactions, encourage them to recall the properties of the surfaces. You can pass around the surface materials again to help them make connections.

Documentation:

Take anecdotal notes about the children's descriptive and observational skills.

Extension Lesson:

This lesson can be extended by moving it to the science center for the children to replicate. It may help to simplify the experiment by using only one liquid at a time with a variety of surface materials. More liquids can be added as children gain experience conducting simple experiments.

Sink or Float Sorting

Topic:
Some objects sink and others float in liquid.

Objective:
Children will test objects that sink and float and will sort objects by the sink-or-float property.

Creativity Skills:

Exploration

Solution finding

Documentation

Opportunities for unique problem solving

Materials:

You can develop various versions of this activity based on the materials you choose to include:

Everyday items (small toys of varying weights and sizes)

Earth materials (leaves, twigs, rocks, pinecones, and seeds of varying weights and sizes)

Ocean/beach items (seashells, rocks, sand, driftwood, sea glass of varying weights and sizes)

Classroom water table or large, shallow plastic bins

Water

Towels (for cleanup)

Science journals and pencils

Overview:

This activity can be done in pairs or small groups either indoors or outdoors.

Activity Steps:

1. To set up, fill the water table or plastic bin with a few inches of water.
2. Talk with the children about the words *sink* and *float*. Encourage them to think about times they may have seen objects sink or float—in the bath with bathtub toys, at the pool with floaties or rafts—and show them the materials they will be exploring today. Do they have an idea which objects will float? Why?
3. Encourage them to test all materials, and give them plenty of time to explore and experiment. Once they are finished exploring, encourage the children to group the objects that floated and those that sank. They can also draw objects that sank and those that floated into their science journals.

Documentation:

The children's groupings of objects and/or journal pages can be used to document their understanding of the sink-or-float property.

Extension Lesson:

This lesson can be extended into the exploration of another property of matter: magnetism. Gather a variety of materials—both metal and nonmetal—for the children to test with magnets. Encourage the children to document their materials tests in their science journals by drawing or writing which objects were attracted to the magnets and which were not.

Describing Color and Movement

Topic:

Matter can be described and classified by its observable properties.

Objective:

Children will participate in this simple experiment and make descriptive observations about the movement of various liquids.

Materials:

One quart-size clear bag for each child (hint: the ziplock-style bags are easiest for children to use independently to ensure a tightly sealed bag)

Water (1/2 cup per child)

Cooking oil (1/2 cup per child)

Cornstarch (5 tablespoons per child)

Food coloring

Creativity Skills:

Exploration

Solution finding

Documentation

Communication/collaboration

Overview:

This lesson has several steps and will work best if a teacher works alongside a small group of children to help them create their color bags and to model and support descriptive language.

Activity Steps:

1. Invite the children to measure and count out 5 tablespoons of cornstarch to add to their bags.

2. Assist children as they add the water and a few drops of food coloring to the bags.

3. Seal the top and mix well.

4. Once the water and cornstarch mixture is combined, open the bag and invite each child to add the oil to the other ingredients.

5. Push as much air out of the bag as possible and reseal.

6. Invite the children to place their bags flat on the table to push the mixture around.

7. Ask questions to promote inquiry: What do you notice about the color bag? Can you describe the color? What happens to the mixture if we let the bags rest for 30 seconds? Why do you think the liquid is separating?

8. If you have accessible windows in the classroom, tape the bags flat against the window, and invite the children to explore the bags. The light from the windows will help illustrate that the mixture is both transparent (oil) and semi-opaque (water/cornstarch) when separated.

Documentation:

Take anecdotal notes about the children's participation and use of descriptive language throughout the experiment.

Extension Lesson:

This lesson can be extended by inviting the children to create oil-and-water discovery bottles to use in the science center. You will need several clean 12- to 18-ounce plastic soda bottles, cooking oil, and colored water. Fill bottles with a ratio of 3/4 water to 1/4 oil. Glue the bottle top on to ensure that it stays closed.

Racing through Liquid

Topic:

Matter can be described and classified by its observable properties.

Different properties are suited to different purposes.

Objective:

Children will participate in a simple experiment and make predictions about the outcome.

Materials:

Canning jars or glass jars of the same size
(1 for each child)

Glass marbles or cabochons
(1 for each child)

Liquids:
    Water
    Light corn syrup

Creativity Skills:

Exploration

Visualization

Communication/ collaboration

Cooking oil
Molasses
Honey

Paper to record results

Pencil

Overview:

This experiment works best in small groups and if there is one child per jar used, because you will be inviting the children to work together to create a fair test. To carry out a fair experiment, the children will need to conduct a fair test. In a science experiment, a *fair test* means that you must change only one factor at a time while keeping all other conditions the same. In this experiment, the type of liquid is the factor that is different, which means that the size of the marble and the time it is dropped must be the same.

Activity Steps:

1. Fill each jar with an equal amount of a liquid.

2. Place the jars next to each other, and invite the children to explore the liquids. You can place a spoon in each jar and invite the children to gently stir each liquid. Which liquid is easiest to stir? Which is hardest?

3. Talk with the children about the experiment; they will be dropping marbles into the liquids to see which marble gets to the bottom first.

4. Have the students practice dropping the marbles on the count of three a few times, before dropping the marbles into the jars. The children will also need to drop the marbles from the same distance, so have them hold their hands over the tops of the jars.

5. Once they are ready, have them drop the marbles and watch carefully to see which marble reached the bottom first and which reached last.

6. Record the results of the first trial.

7. Conduct two more trials to see if the results are the same. If all factors were kept consistent, the results should be the same. If the results vary, invite the children to think about whether they are conducting a true fair test.

8. Repeat trials as needed or desired. Be sure to talk with the children about why the marbles moved through some liquids faster than others. Connect the experiment results back to what the children experienced while stirring the liquids.

Documentation:

Informal observations of the children's abilities to conduct a fair test and their descriptions of the liquids will help you understand their thinking about the properties of liquids.

Extension Lesson:

This lesson can be extended by using different liquids or different objects to drop into the liquids. Encourage the children to think about how the properties of the liquids and the objects affect the results of their trials.

Children's Books

Beaty, Andrea. 2016. *Ada Twist, Scientist.* New York: Abrams Books for Young Readers.

Ada Twist, Scientist follows in the footsteps of *Rosie Revere, Engineer,* and *Iggy Peck, Architect* with an endearing story of one girl's persistence at solving science mysteries. Your students will relate to Ada's mischievous explorations and be inspired by her thoughtful inquiries into the world around her.

Ross, Michael Elsohn. 2007. *What's the Matter in Mr. Whiskers' Room?* Somerville, MA: Candlewick Press.

Your students can join Mr. Whiskers's class as they spend the day exploring science centers focused on matter in the classroom and on the playground.

Schuh, Mari. 2011. *All about Matter.* North Mankato, MN: Capstone Press.

This book looks into the three states of matter: solids, liquids, and gases. Colorful photographs and text that connect to children's everyday lives provide an engaging entry point for young children.

2
Physical Science: Physical and Chemical Changes

Building Understanding of Physical and Chemical Changes

One spring day, I was observing a class of kindergarten students as they watched their teacher drop a Mentos candy into a 2-liter bottle of carbonated cola. As the cola shot into the air, the children cheered excitedly. Once back in the classroom, I asked a student why he thought the cola bubbled and fizzed out the top of the bottle. He replied with one word: "Magic!" One of the most endearing characteristics of young children is their willingness to believe in magic. Believing in magic allows children to formulate explanations for ideas and concepts they cannot quite understand yet. Many physical-science experiences involve changes that are hard for children to explain and understand.

Teachers must recognize that young children will rely on this type of explanation, so it is important to talk through children's explanations of events like the Mentos/cola experiment. Having children watch an exciting experiment, while fun, will not do much to build their understanding of physical science. To build on children's experiences, provide opportunities for your students to experiment with familiar materials. Children will be able to explore what happens when certain substances are combined and will observe the reactions. Working in a hands-on, minds-on manner will require the children to engage, explore, explain, extend, and evaluate their thinking and experiences.

Understanding Physical and Chemical Changes

Matter is anything that has mass and takes up space. Matter is made up of tiny particles called *atoms* and *molecules*. Atoms and molecules make up the three common states of matter: solids, liquids, and gases. A *solid* can hold its own shape. The particles in most solids are packed closely together, and solids are hard to compress (imagine trying to squash a wood block with your hands). A *liquid* can be poured from one container into another and takes the shape of the container it is in. Most liquids have particles that are less densely packed than solids, which means that they can move past each other. While solids are very difficult to compress, liquids are also difficult to compress (imagine what would happen if you attempted to compress a liquid within in a cup). The atoms and molecules in gases are much more spread out than in solids or liquids and they can move freely. A *gas* will fill any container and can be easily compressed.

Matter is classified based on its chemical and physical properties. *Physical properties* are characteristics that can be observed or measured without changing the composition of a substance (size, shape, color, texture, hardness). A *chemical property* may only be observed by changing the chemical identity of a substance. *Chemical changes* occur when two substances are mixed together and form something new. This differs from a *physical change*, which is a substance that changes physical forms but still retains its original properties—water frozen to ice can return to its original state, so it is only a physical change. There are four clues that can help you determine whether a chemical change has occurred:

- There is a formation of gas, evidenced by fizzing and bubbling
- The reaction causes heat, light, or odor
- A color change is produced
- The reaction causes a solid to form

The Cycle of Inquiry: Physical and Chemical Changes

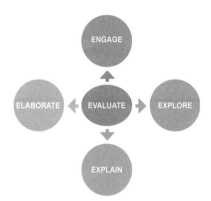

Engage

Questions to Engage: Physical and Chemical Changes

Ask questions such as the following to engage young learners' interests:

- What do you think will happen when we mix these two substances? Why?
- Have you ever seen a substance like this before?
- What do you think this substance feels like? smells like?
- Can you describe this substance with one word?

Explore

Engaged Exploration: Physical and Chemical Changes

To assist children as they work to explore the world around them, provide a space in the classroom that invites them to do the following:

- Watch and participate in simple experiments
- Touch, smell, and manipulate substances under your supervision
- Draw or write their observations of the results of experiments

Explain

Explanation Opportunities: Physical and Chemical Changes

Use the following suggestions to help children understand and explore class lessons on a deeper level:

- Display photo charts of solids, liquids, and gases with the properties listed

- Display photographs of experiments with student quotes and trial results

- Plan whole-group debriefings that discuss observed physical or chemical changes

- Encourage students to write, draw, and comment in their science journals

- Display whole-class exploration charts based on substances that melt or don't melt and substances that mix together or those that do not

Elaborate

Science Center Elaborations: Physical and Chemical Changes

Keep your students engaged with the following ideas to continue learning in the classroom:

- Make simple science bottles (water and oil, sand and water, water and glitter)

- Repeat experiments based on interests

- Classify materials based on the properties of solids, liquids, and gases

Physical Science: Physical and Chemical Changes

Evaluate

Documenting Informal Evaluation: Physical and Chemical Changes

Help your students evaluate their learning from the explorations by using these suggested ideas:

- Plan whole-class review sessions
- Have a science night for families and friends
- Display documentation panels of major projects
- Display student portfolios
- Involve students in lesson planning based on their thoughts and ideas

Core Ideas in Physical and Chemical Changes

Properties of Matter

Different kinds of matter exist, and many of them can be either solid or liquid, depending on temperature. Matter can be described and classified by its observable properties.

Different properties are suited to different purposes.

Properties of Chemical Changes

Heating or cooling a substance may cause changes that can be observed. Sometimes these changes are reversible, and sometimes they are not.

Source: NGSS Lead States. 2013. *Next Generation Science Standards: For States, by States.* Washington, DC: National Academies Press.

Vignette for Understanding: Growing Crystals

Ms. Sandy is introducing her class to an integrated science and visual-arts project where the children will be working together to create their own crystals. This project will take several days and will allow the children opportunities to observe changes over time.

The class is gathered together around a large tray where Ms. Sandy has placed four glass jars. She shows the children the materials they will use to create their crystals—hot water, food coloring, and a special kind of salt called Epsom salts. She passes a piece of Epsom salt around so the children can make comparisons between it and table salt. The children note that Epsom salt is much larger. Ms. Sandy asks for four volunteers to squeeze a few drops of food coloring into each jar. Once that is done, Ms. Sandy carefully pours 1/2 cup of hot water into each jar, and the children watch as the colors quickly mix. Ms. Sandy invites four children to add 1/2 cup of Epsom salts to each jar. The children take turns stirring until the Epsom salts are mostly dissolved. She invites the children to come up one by one and take a careful look into each jar. She takes a photograph of the inside of each jar to use over  the next few days to help the children make comparisons of the crystal growth over time. Ms. Sandy places the jars in the refrigerator and lets the class know that they will make

Physical Science: Physical and Chemical Changes

observations about the growth of the crystals every morning over the course of the week. The next day, the class takes a close look and sees a network of interconnected crystals growing!

Reflection

Ms. Sandy was able to involve her students in the creation of the crystal jars during the daily observations. When doing longer-term projects with young children, it is important to have documentation, such as the photographs of the jars that Ms. Sandy took, to remind the children of previous observations. The documentation photographs can be used each day to help children make comparisons and predictions about the possibility for growth over the next day. The inquiry process in this experience includes opportunities for students to make observations, explore, question, make predictions, use simple tools, and conduct simple science investigations.

Matter and Physical Properties across the Curriculum: Planning Tips

Experiences with physical and chemical changes will likely take place with your close supervision in small- and large-group experiences. Because many of these experiments will require adult supervision and participation, you may encourage parent interest by asking them to re-create the activities you do in your classroom at home. Providing parents with step-by-step instructions and tips based on the children's earlier experiences will help support parents as they work with their children at home. This extension between school and home will also encourage your students to share what they know with their families. Be sure to follow up with parents and caregivers to find out the results of their at-home experiments and invite them to ask questions about what the children are learning as they work in the sciences.

Lesson Ideas

Gooey Explorations

Topic:

Different kinds of matter exist, and many of them can be either solid or liquid, depending on temperature. Matter can be described and classified by its observable properties.

Objective:

Children will participate in measuring, mixing, and manipulating a variety of playdough/clay substances.

Materials:

No-Cook Playdough

2 cups plain flour

1 cup salt

1 tablespoon oil

1 cup cold water

2 drops food coloring

Fizzy Dough

For this dough, you can make a large batch with one color or several smaller batches of different colors. The following materials will make a small batch:

1/2 cup flour

1/2 cup baking soda

2 tablespoons oil

> **Creativity Skills:**
>
> Exploration
>
> Strategic planning
>
> Documentation
>
> Communication/ collaboration
>
> Opportunities for unique problem solving

3 to 5 drops of food coloring

1/4 to 1/8 cup vinegar

Powdered Drink Mix Playdough

2 cups flour

1 cup salt

1 package powdered drink mix (sugar-free)

1 cup hot water

Overview:

These playdough recipes allow for children's active involvement because they do not involve any microwave or stove cooking. Invite children to work alongside you in pairs or small groups as you make the playdough. Encourage the children to measure and mix the ingredients. As ingredients are added, you can ask prompting questions such as: What do you think will happen when we add this ingredient? How does the dough look now? What does it smell like?

To Make No-Cook Playdough:

1. Combine flour and salt.
2. Add water, food coloring, and oil, and mix until combined.
3. Take dough out of the mixing bowl and knead until soft. If the dough is too soft, you can add a little more flour (one teaspoon at a time).

To Make Fizzy Playdough:

1. Begin by mixing the oil and food coloring in a small bowl until combined.
2. Add the flour and mix thoroughly.
3. Add the baking soda, and mix until combined (it will be a coarse mixture).

4. Slowly add the vinegar over the top of the flour mixture. It can be poured from a measuring cup or put into a squirt bottle, which will be easier for the children to manipulate. Once the chemical reaction has occurred, the mixture can be kneaded by hand and will have the playdough-like texture, though it will be lighter.

To Make Powdered Drink Mix Playdough:

1. Add dry ingredients to a bowl and mix until combined.

2. Slowly add the hot water and carefully mix until combined.

3. Take dough out of the mixing bowl and knead until soft. If the dough is too soft, you can add a little more flour (one teaspoon at a time).

4. Once the children have made the dough, be sure to ask them to describe what they are seeing and smelling and how the dough feels.

Documentation:

You can make informal observations of children's abilities to follow directions and work collaboratively as you work alongside them. You can also note the children's responses to your questions and observations to plan follow-up experiences for the class.

Extension Lesson:

Once the playdough is made, place it on a sand/water table or on small tables for children to explore. This lesson can also be extended by developing more step-by-step experiences (an important part of experimentation in science) in your classroom. Simple cooking experiences, such as making sugar cookies, will allow children to follow a set of directions, work collaboratively, and observe physical reactions of matter.

Bouncy Ball Science

Topic:

Different kinds of matter exist, and many of them can be either solid or liquid, depending on temperature. Matter can be described and classified by its observable properties.

Objective:

Children will participate in measuring and mixing the substances to create their own bouncy ball.

Materials:

3 tablespoons of cornstarch (per ball)

Water

Food coloring

Measuring cups (wet ingredients)

Measuring spoons (dry ingredients)

Small, microwave-safe cups or bowls for mixing

Spoons for stirring

Plastic bags for storage

Microwave

Creativity Skills:

Exploration

Communication/ collaboration

Opportunities for unique problem solving

Overview:

This experiment has many steps and will require your participation and supervision throughout. It will work best to do this with children in pairs or small groups.

Activity Steps:

1. Ask the children if they've ever played with a bouncy ball. Can they describe how the ball moves? Invite them to join you in making their own ball in this experiment.

2. Ask the children to choose a color for their ball.

3. Measure 3 tablespoons of cornstarch, 1 tablespoon and 1 teaspoon of water, and 2–3 drops of food coloring and place in mixing bowl.

4. Invite a child to stir the mixture until it is well mixed. The mixture will be difficult to stir.

5. Microwave the mixture for 20 seconds on regular power setting.

6. Remove the mixture from the microwave. It should be warm but not hot.

7. Stir in 1 1/2 teaspoons of water. Mix well. The mixture should be easy to collect.

8. Encourage a child to roll the mixture in his hands slowly to carefully shape a ball. If the mixture is too dry and the ball has cracks, brush a small amount of water onto the cracks and allow it to rest for a minute before reshaping.

9. As a last step and to help the ball retain its shape, microwave the shaped ball for 15 seconds.

10. Repeat the process for the remaining children in your group. The balls will bounce when gently dropped onto a table, but discourage the children from throwing the balls. The balls will remain soft for some time if stored in a plastic bag but will need to be re-formed because it will lose its shape when it rests for a period of time.

Documentation:

You can make informal observations of children's abilities to follow directions and work collaboratively. You can also note the children's responses to your questions and make observations to plan follow-up experiences for the class.

Physical Science: Physical and Chemical Changes

Extension Lesson:

This lesson can be extended by encouraging the children to interact with other simple substances (such as those in the Gooey Explorations lesson) that go through physical or chemical changes when mixed.

Color Races

Topic:

Heating or cooling a substance may cause changes that can be observed.

Objective:

The children will participate in a simple experiment and notice the differences occurring among warm, hot, and cool liquids.

Materials:

3 clear glasses

1/2 cup each of hot, room-temperature, and cold water

Food coloring (choose one color for all three water temperatures)

Timing device that counts seconds

Creativity Skills:

Exploration

Solution finding

Communication/collaboration

Overview:

This experiment will work best in a whole-group setting. Students will learn about the effect of water temperature on molecules.

Activity Steps:

1. Talk with the students about how water is made up of tiny particles known as molecules that we can't see but we know are there. The molecules in water are moving all the time. Some water temperatures have molecules that move faster than at other temperatures. Let them know that today they will be helping with

an experiment to see which temperature of water has the fastest-moving molecules.

2. Fill a clear glass with some room-temperature water.

3. Invite the students to predict how long (number of seconds, minutes) they think it will take for all of the water in the glass to change color when you add food coloring to it.

4. Add a few drops of food coloring to the water, and time how long it takes for the color to completely disperse. Invite the children to count along with you.

5. It is important not to move or stir the cup while you are watching the color disperse.

6. Follow the same procedures with hot water, and invite the children to make predictions about whether the hot water will change faster or slower than room-temperature water.

7. Note the time difference between the liquids, and ask them to think about why the hot water may have changed quicker.

8. Follow up with the cold water. Ask children to predict whether the cold water will change faster or slower than the room-temperature or hot water. The cold water will take longer, so be prepared to count minutes rather than seconds.

9. Once the color in the cold water has dispersed, you can redo the experiment with all three water temperatures at the same time so that the children can watch and compare the reactions.

10. End the experiment by asking the children to think about why the hot water always changed color fastest. (Hint: The faster the liquid molecules move, the less time it takes for the food coloring to dilute. The molecules help stir the food coloring around.)

Documentation:

Your informal observations of the children's predictions and explanations during the experiment trials will help you understand their learning.

Extension Lesson:

This lesson can be extended by trying out the experiment with another liquid, such as milk or a clear juice, to see if the type of liquid makes a difference.

Moving Molecules Game

Topic:

Heating or cooling a substance may cause changes that can be observed.

Objective:

The children will participate in a game to reinforce the idea that the molecules in warm, hot, and cool liquids behave differently.

Materials:

None needed

Overview:

This game works best when you play it after you've completed the Color Races experiment. Play as a whole class in an area of the classroom where the children have room to move around.

Creativity Skills:

Exploration

Communication/collaboration

Opportunities for unique problem solving

Activity Steps:

1. Have the whole class stand up, and explain that they are now going to "become" molecules. Explain that warm molecules move around a lot and like to spread out away from each other. Ask students to do this by walking, bouncing, or dancing. The hotter the molecules are, the faster they move.

2. Invite the students to act like hot molecules.

3. Next, they are going to become cold molecules. Remind the students that cold molecules like to huddle together and bounce slowly. The colder the molecules are, the slower they bounce. When they freeze they bounce really slowly, but they don't stop moving.

4. Invite the students to act like frozen molecules.

5. Give students cues as to how they should act. Call out hot molecules, cold molecules, and warm molecules, and students should act accordingly. Start off changing cues slowly but then get faster.

Documentation:

Your informal observations of the children's actions during the game will help you understand their thinking about molecules and substance temperature.

Extension Lesson:

You can extend the game by playing as a solid (huddle very close together), a liquid (spread out a bit but stay in a confined area), and as a gas (spread out everywhere).

Balloon Blow-up

Topic:

Combining certain substances can cause changes that can be observed. Sometimes these changes are reversible, and sometimes they are not.

Objective:

Children will participate in measuring and mixing substances and observing the changes that happen.

Materials:

Balloon (1 per trial)

Small bottle (cleaned 16-ounce soda bottle)

Small funnel (if you don't have one, you can make one out of heavy paper for this experiment)

Baking soda (2 tablespoons per trial)

Creativity Skills:

Exploration

Documentation

Communication/ collaboration

Opportunities for unique problem solving

Physical Science: Physical and Chemical Changes

Vinegar (4 ounces per trial)

Droppers/pipettes (optional)

Paint (optional)

Overview:

Invite the students to do this experiment with you, with one pair of students at a time. This is a fast-moving experiment with several precise steps.

Activity Steps:

1. Ask the students if they've ever blown up a balloon or seen someone blow up a balloon. Ask them what makes the balloon expand. Let them know that today they will help blow up a balloon but they won't be blowing air from their mouths into the balloon. They will be mixing vinegar and baking soda to see if they can blow up the balloon.

2. Ask them to think about what will happen when baking soda and vinegar come in contact that may help inflate the balloon.

3. Using the funnel, add the baking soda to each balloon (two people may be needed for this: one person to hold the balloon open and the other person to put the baking soda inside of the balloon).

4. Pour the vinegar into the bottle using the funnel.

5. Carefully fit the balloon over the bottle opening, taking care not to drop the baking soda into the vinegar yet.

6. Once the balloon is fit tightly on the nozzle, hold up the balloon and allow the baking soda to fall into the vinegar.

7. Observe the chemical reaction and effect on the balloon.

8. Ask the children to think about what is produced that is blowing up the balloon. Introduce them to the term *gas,* and repeat the experiment to reinforce their ideas and observations. A great follow-up discussion question is to invite them to think about why the balloon eventually stops blowing up. Draw their attention to the liquid at the bottom—do they still see fizzing, or has that stopped?

Documentation:

Your informal observations of the children's actions during the game will help you understand their thinking about gases.

Extension Lesson:

This lesson can be extended by exploring other fizzing experiments. For the Fizzing Colors experiment, you will add small amounts of food coloring to baking soda and mix until combined. Using droppers/pipettes, invite the children to add vinegar to the baking-soda mixture. The children can paint with the mixture once the colors are created and the fizzing has stopped.

Children's Books

Bradley, Kimberly Brubaker. 2001. *Pop! A Book about Bubbles.* New York: HarperCollins.

Illustrated with colorful photographs, *Pop! A Book about Bubbles* explores how bubbles are made, why they float, and what makes them pop. The book ends with instructions for creating your own bubble solution.

Carle, Eric. 1992. *Pancakes, Pancakes!* New York: Simon & Schuster.

As Jack and his mother cook pancakes for breakfast, they learn about the physical changes (cracking an egg open and grinding the wheat into flour) and chemical changes (burning wood for cooking, cooking the batter to make pancakes) happening while they cook.

Dr. Seuss. 1949. *Bartholomew and the Oobleck.* New York: Random House.

A Caldecott Medal winner, *Bartholomew and the Oobleck* differs from many Dr. Seuss classics in that it is written in prose. This is a longer book that you can read over the course of several days with your students. Children will enjoy this tale about the creation of oobleck, a messy green substance that covers Bartholomew's town. Following this story, you can make oobleck with your class using this simple recipe.

Oobleck

1 cup of water
1 to 2 cups of cornstarch
Mixing bowl
Green food coloring (3–5 drops)
Spoon

Activity Steps:

1. Working with a small or large group, invite students to take turns completing each of the following steps.

2. Mix the food coloring into water.

3. Pour one cup of cornstarch into a bowl. Slowly add the food-coloring water, mixing as you go until the mixture is firm.

4. Add more cornstarch if the mixture is too runny or more water if it is too thick. Pour the water in, mixing slowly as you go.

5. Keep adding more water until the mixture becomes thick and hardens when you tap on it.

6. Add more cornstarch if it gets too runny and more water if it becomes too thin.

Tip: Oobleck is a non-Newtonian fluid, which means that it is pressure dependent. If you apply pressure, the mixture feels hard. If you remove pressure, the mixture softens.

3
Life Science: Growth and Change

Life Science Experiences that Build Knowledge of Growth and Change

Growth and change in life science are important topics of natural interest to young children. Young children enjoy watching plants and animals grow and change. Anyone who has ever brought up the topic of puppies and kittens in a preschool class can share just how exciting that topic is for young children. Animal babies are a natural entry point for teachers to introduce ideas about how baby animals grow into adults. We can build on this natural curiosity about living things to expand children's everyday observations.

Understanding Growth and Change

Plants have basic needs—air, water, light, and appropriate temperatures—that must be met for them to survive and grow into maturity. Plant structures—roots, leaves, stems, flowers, and seeds—have different roles and contribute to the ability of a plant to survive. Plant roots bring in water and nutrients (food) from the soil. Plant stems hold the plant upright and contain xylem and phloem (transport tissues) to deliver the water and nutrients brought in by the roots. Plant flowers are the reproductive center of the plant. Plant leaves absorb sunlight and exchange gases—oxygen for respiration and carbon dioxide for photosynthesis. Photosynthesis is a process that uses sunlight to synthesize foods from carbon dioxide and water. Just like plants, animals have basic needs for survival—air, water, food, and safe shelter. Animals, like plants, are living things that grow and change. Animals, including humans, have offspring that closely resemble their parents and need parental protection to grow. Animals are affected by the environment, and their behaviors are influenced by the conditions in their environment (for example, birds migrating to warmer climates in the winter to survive).

The Cycle of Inquiry: Growth and Change

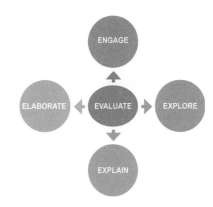

Engage

Questions to Engage: Growth and Change

Ask questions such as the following to engage young learners' interests:

- What do you know about this plant? Can you show us where the stem (leaves, flowers) is located?
- Have you ever seen a seed like this before?
- What do you think a baby chick feels like? What would the chick sound like?
- How could you describe what a tree looks like to someone who has never seen one?

Explore

Engaged Exploration: Growth and Change

To assist children as they work to explore the world around them, provide a space in the classroom that invites children to do the following:

- Plant, monitor, and take care of a variety of plants (indoors or outdoors)

Life Science: Growth and Change

- Explore the different parts of flowers and plants
- Draw or write their observations of plant growth
- Explore photos and books about different kinds of animals

Explain

Explanation Opportunities: Growth and Change

Use the following suggestions to help children understand and explore class lessons on a deeper level:

- Display photographs of plant growth and change
- Plan whole-group debriefings that discuss observed changes over time
- Encourage students to draw, write, and comment in their science journals
- Display charts that match baby animals to their parents. These can also include information about habitats and needs.

Elaborate

Science Center Elaborations: Growth and Change

Keep your students engaged with the following ideas to continue learning in the classroom:

- Plan seed explorations
- Observe plant growth daily
- Use measuring tools to track the growth of students over the year

Evaluate

Documenting Informal Evaluation: Growth and Change

Help your students evaluate their learning from the explorations by using the suggested ideas:

- Plan whole-class review sessions
- Have a science night for families and friends
- Display documentation panels of major projects
- Display student portfolios
- Involve students in lesson planning based on their thoughts and ideas

Core Ideas in Life Science: Growth and Change

Understanding Ecosystems

- All animals need food to live and grow. They obtain their food from plants or from other animals. Plants need water and light to live and grow.

- Living things need water, air, and resources from the land, and they live in places that have the things they need. Humans use natural resources for everything they do.

Source: NGSS Lead States. 2013. *Next Generation Science Standards: For States, by States.* Washington, DC: National Academies Press.

Vignette for Understanding: Exploring Plant Roots

Each child in Cullen's class planted marigold seeds about eight weeks ago, and the class has been carefully monitoring each plant's growth. Cullen has been drawing his observations each week in his plant journal. After a couple of weeks of drawing the marigold flowers, Cullen asked his teacher, Ms. Spencer, if he can see the plants' roots. Ms. Spencer and Cullen talk about how delicate plants are and what each part of the plant does. Ms. Spencer asks Cullen if he is okay with taking his plant out to examine its various parts, knowing that he won't be able to put it back together. Cullen decides that he's ready to do so to look at the roots and touch them. He wonders with excitement what the roots will feel like and decides that "they will probably feel like hair." He calls over a friend, Marcus, to the table and excitedly shares the news that Ms. Spencer is going to help him take his

plant out of the pot so that he can see and feel the roots. Ms. Spencer comes back to the science center with a large tray and a few wooden skewers. She asks Cullen to hold the marigold pot while she carefully pulls the plant loose. The boys both make an "ohhh" sound as the plant comes loose from the pot with roots and soil intact. Ms. Spencer lays the plant on the tray and demonstrates how the boys can carefully loosen the soil from around the roots with the skewers.

The boys work together to loosen the soil, and as they work, Cullen carefully touches the roots, noting that they aren't slimy as he thought they might be. He declares that they feel like "flower roots" and not like anything else. Ms. Spencer comes to view their progress and shares small magnifying glasses with the boys so that they can take a closer look. Ms. Spencer reminds Cullen to draw the roots in his plant journal before he leaves the center. Other children come and join the boys to feel the roots and help Cullen loosen the dirt while he takes time to document his plant one last time.

Reflection

Cullen was able to watch a plant grow from a seed to a flowering state over the course of many weeks. This type of firsthand experience is incredibly valuable because with each passing week, Cullen's knowledge of a plant's needs and parts were deepened through his observations. Ms. Spencer made an important decision to encourage the children to document their observations each week in their science journals. This will allow the children to look back and track the plant's growth over time. Ms. Spencer was also open to Cullen's idea to unpot his flower and explore the roots. This allowed Cullen, Marcus, and the other children to make firsthand observations about plant roots, extending their learning over the course of the project. The inquiry process in this experience included opportunities for students to make observations, explore, question, make predictions, use science tools, and conduct simple science investigations.

Growth and Change across the Curriculum: Planning Tips

Experiences with growth and change provide natural connections to mathematics. As children measure plant growth and sort seeds, they use measurement and classification skills. Finding ways to integrate mathematics experiences can be very valuable and helps deepen your students' experiences. Encouraging children to record their measurements and the results of sorting and classification experiences in their science journals will provide an opportunity for them to track their learning and make their understanding visible to themselves and others. Remember to give children ample time to write or draw their understanding—documentation is an important part of the inquiry cycle and will help children understand how to share their ideas with others.

Lesson Ideas

Seed Exploration and Matching

Topic:

Plants grow from seeds. Each seed will only grow a specific type of plant.

Objective:

Children will explore a variety of seeds and the associated fruit or vegetable.

Materials:

A variety of fruits and vegetables that can be cut open and explored by children. **Safety note:** Keep in mind the ages and developmental levels of your students when choosing seeds to explore to avoid instances of accidental (or intentional) ingestion. Also, be aware of any allergies or other concerns. Fruits and vegetables with seeds of varying sizes include grapes

Creativity Skills:

Exploration

Communication/collaboration

with seeds, apples, oranges, small pumpkins or gourds, pomegranates, strawberries, and avocados

Magnifying glasses

Small bowls of water

Small spoons

Paper or cloth towels

A small bowl for each type of fruit or vegetable available for exploration

Cutting board and knife (adult use only)

Digital camera for documentation

Science journals and pencils

Overview:

This is an exploration activity that will work best in your science center or at a table where children can work in pairs or small groups, though you can introduce the activity to the whole group. Limit the number of fruits/vegetables to explore if this is the first time your students will have this type of experience. You can even choose to explore one fruit or vegetable per day for a week or more.

Activity Steps:

1. Begin by passing around the uncut fruits and vegetables, and invite children to name them and make predictions about what the seeds might look like.

2. Ask children prompting questions—Do you think this pumpkin has big seeds? What color will they be? Why do you think that? Have you seen the inside of a pomegranate before?

3. Once children have explored each fruit, cut the fruit open and show the children. Were their predictions correct?

Life Science: Growth and Change

4. Let the children know that the fruits and vegetables will be placed in the science center for them to explore further. Place the cut fruits, small bowls of water (to clean seeds), paper or cloth towels (to dry seeds and hands), small spoons (for scooping seeds), empty bowls (for seed collection), and magnifying glasses in the science center.

5. Invite the children to collect the seeds from each fruit, clean them, and place the seeds in bowls. Encourage the children to place the seeds from each fruit in a different bowl so they can be dried and explored further by others. Children can use their science journals to draw representations of the seeds and fruits.

Documentation:

Take anecdotal notes about the children's observations. Science journal drawings can also be used to document the experience and student thinking.

Extension Lesson:

This lesson can be extended by keeping the seeds in the science center for further exploration. A photo of the fruit can be placed beside the seeds to remind students of the connections between the seeds and fruit.

Understanding Roots: The Celery Experiment

Topic:

Plants need water to live, and the roots help bring water to the plant.

Objective:

Children will participate in the setup and observation of a simple experiment to explore how roots bring water to plants.

Materials:

Celery stalks with leaves attached (you will need 1 or 2 stalks for each color you use in the experiment)

Food coloring

Water

Clear glass or heavy plastic jars or vases (1 jar per color used in the experiment. Two or more colors are recommended so that children can explore observable differences.)

Science journals

Colored pencils and markers

Scissors (adult use)

Digital camera (optional)

> **Creativity Skills:**
>
> Exploration
>
> Visualization
>
> Documentation
>
> Communication/ collaboration

Overview:

This is a classic science experiment that will take place over three days in your classroom. You can conduct the experiment in a whole- or small-group setting, but be sure to place the jars in an area of the classroom where the children can make observations without disturbing the celery jars.

Activity Steps:

1. Prior to the lesson, cut several stalks with leaves from the main celery bulb. If the stalks are already cut from the bulb, trim about a 1/2 inch off the bottom of each stalk.

2. Invite the children to pass around and explore the celery stalks. If allergies or food restrictions are not an issue, invite the children to taste the precut celery samples.

Life Science: Growth and Change

3. Talk with the children for a moment about how plants get water. Invite them to tell you what they know.

4. Pour about a cup of water into the clear jar or vase.

5. Invite a child to squeeze four drops of food coloring into the jar and carefully stir the water until the food coloring is evenly dispersed. Repeat this for each color you would like to use for the experiment. Tip: Yellow and green will be harder for the children to detect on the celery stalks and leaves.

6. Invite the children to place one or two stalks in each jar.

7. Talk with the children about their predictions about what will happen to the stalks and leaves. Encourage them to explain their thinking with prompts such as, "Can you tell us why you think that will happen? Why do you think that?"

8. Invite the children to make observations at timed intervals the first day of the experiment. Checking every three hours will allow them to explore any immediate results. Significant results will be seen after the first 24 hours and then again at 48 hours. Encourage the children to draw or write their observations each time in their science journals so that they can reflect on earlier observations to help note any changes. You can also take photographs to print out later and put into the science center to remind children of their experiment results.

Documentation:

Children's science journals will provide documentation of their observations and understanding of changing colors.

Extension Lesson:

This lesson can be extended by bringing the celery into the science center for close observation after the leaves have changed colors. You can cut the celery stems both vertically and horizontally so that the children can see how the water traveled up the stalk to reach the leaves.

Plant Growth Journal: From Seed to Flower

Topic:
Plants need water, nutrients, and light to live and grow.

Objective:
Children will carefully observe and document the life cycle of a plant.

Materials:

Seeds (seeds that grow well indoors quickly include lima beans, zucchini, radish, zinnia, lettuce, marigold, and basil)

Soil

Clear plastic cups

Plant journal with enough pages for children to complete 2 observations a week for 4 to 6 weeks

> **Creativity Skills:**
>
> Exploration
>
> Visualization
>
> Documentation
>
> Communication/collaboration

Optional materials/media to assist in documentation (a variety of markers, crayons, and colored pencils, digital cameras, magnifying glasses, ruler or nonstandard measurement tool, and a color wheel to assist in naming colors)

Overview:

This is a longer project, so take the necessary time to explain to children what they will be doing over the following weeks.

Activity Steps:

1. Begin by inviting the children to explore the seeds they will be planting. They can choose a plant type to grow and begin the process by documenting what their chosen seed looks like.

2. Demonstrate the planting process and invite the children to plant their seeds (typically planting at least 3 seeds per cup will result in at least 1 successful germination).

3. Place seeds in an area of the classroom where the cups will receive light and where children can make their biweekly observations without disturbing the cups too much.

4. Be sure to invite children to share their observations over the course of the project to keep their interest.

SAMPLE PLANT JOURNAL PAGE

Plant Project

Name: _____   Date: _____

_____ My plant is below the ground today
_____ My plant has emerged!   It is _____ tall   It is _____ color

My observation:	What I think I will see next:

Life Science: Growth and Change

Documentation:

The children's journal pages can serve as documentation of the learning process, and so can your informal observations from the children's discussions and observations over the course of the project.

Extension Lesson:

This lesson can be extended by inviting the children to plant their flowers in a class garden, take them home for planting, or take them apart for closer examination.

Babies and Their Moms: Matching Game

Topic:

Baby animals have characteristics similar to their parents.

Objective:

Children will explore a matching game that asks them to match a baby animal with its parent.

Materials:

3" x 5" or 4" x 6" photos of animal babies and their parents (a minimum of 10 to match one-to-one)

Overview:

This game can be played individually or in pairs/small groups. Children will be matching photos to learn about animals. If this is the children's first time playing this game, it is helpful to do a few practice rounds together.

Creativity Skills:

Exploration

Solution finding

Visualization

Strategic planning

Communication/collaboration

Activity Steps:

1. Before beginning play, talk with the children about matching objects. In particular, focus on how the game will ask them to match one baby animal with its parent. Some questions to consider are the following:

What animal baby is this? How do you know that? What is similar about the way the baby looks when you compare her to her mom?

2. Invite the children to place all of the cards facedown.

3. Invite a child to turn over one card, followed by another card to try to create a match.

4. If no match is made, turn both cards over and invite the next child to take a turn. Make sure the children have space to lay out their one-to-one matches.

Documentation:

You can use informal observations of the children's physical matches as documentation of their experience.

Extension Lesson:

This matching game can be modified to invite children to match an animal with its habitat, such as a bird with a nest, a fish with water, and so on.

Children's Books

Ehlert, Lois. 2004. *Pie in the Sky*. Boston, MA: HMH Books for Young Readers.

Your students can follow along as a father and son observe a cherry tree in their backyard over the four seasons. The colorful collage illustrations show the beauty of each season.

Ghigna, Charles. 2012. *Little Seeds*. Bloomington, MN: Picture Window Books.

Little Seeds is a beautifully illustrated poem that shares how planting seeds is good for the Earth.

Michels, Dia. 2005. *If My Mom Were a Platypus: Mammal Babies and Their Mothers*. Washington, DC: Platypus Media.

This book shares the stories of 14 mammal baby and mother pairs (including humans) as the mothers feed, guide, protect, and teach their babies. The story of each mammal is very detailed, so you may find that your students enjoy hearing about one baby and mother pair at a time.

Wheeler, Eliza. 2013. *Miss Maple's Seeds*. New York: Nancy Paulsen Books.

Miss Maple's Seeds shares a nurturing narrative about a woman who gathers and cares for lost seeds until spring, when it is time for them to grow. The book features colorful paintings and a page that realistically depicts 20 seeds, from raspberries to acorns.

4
Earth Science: Conservation and Sustainability

Fostering Interest in Conservation and Sustainability

Young children take a strong interest in taking care of the environment, so the early childhood years are an ideal time to introduce terms such as *reduce, reuse, recycle, conservation,* and *sustainability.* You can weave these concepts throughout the everyday life of your classroom—taking only what you need, reusing paper, and recycling plastic waste—and through planned experiences such as the ones described in this chapter. Caring for the environment will involve modeling and encouraging appropriate behaviors as well as talking with children about what they can do to help.

Understanding Conservation and Sustainability

The three *R*'s of conservation and sustainability are action steps that we can take: reduce, reuse, and recycle. *Reducing* is the act of cutting back the amount of trash we generate. *Reusing* is the act of finding new ways to use waste so that it doesn't end up as trash. *Recycling* is the act of using trash to make new objects or goods. Recycling is a process known as the recycling loop: the loop begins when we take items to be recycled to recycling centers or recycling bins; the items are then taken to a processing center where they can be sorted and processed into raw materials; the raw materials are sold to manufacturers that make new items with them; and the loop is complete when people use items that were made from the recycled materials. *Sustainability* ensures that the needs we have in the present day do not compromise the ability of future generations to meet their own needs.

The Cycle of Inquiry: Conservation and Sustainability

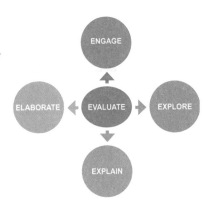

Engage

Questions to Engage: Conservation and Sustainability

Ask questions such as the following to engage young learners' interests:

- Can this object be recycled? How do you know that?
- How can we reuse this object?
- What's another way we can use this object?
- Let's think about ways that we can reduce the amount of paper we throw away.

Explore

Engaged Exploration: Conservation and Sustainability

To assist children as they work to explore the world around them, provide a space in the classroom that invites children to do the following:

- Collect materials that can be reused later in different ways
- Recycle paper, plastic, and glass items that can be taken to a recycling center

Explain

Explanation Opportunities: Conservation and Sustainability

Use the following suggestions to help children understand and explore class lessons on a deeper level:

- Display charts of materials that can be recycled
- Display photographs of recycling walks or efforts to clean the playground of trash
- Plan whole-group discussions of recycling efforts
- Encourage children to draw, write, and comment in their science journals

Elaborate

Science Center Elaborations: Conservation and Sustainability

Keep your students engaged with the following ideas to continue learning in the classroom:

- Sort and classify materials into items for trash, items for reuse, and items for recycling
- Plan reuse experiences that create new artworks or play items from discarded objects

Evaluate

Documenting Informal Evaluation: Conservation and Sustainability

Help your students evaluate their learning from the explorations by using these suggested ideas:

- Plan whole-class review sessions

- Have a science night for families and friends
- Display documentation panels of major projects
- Display student portfolios
- Involve students in the planning of next steps based on their thoughts and ideas

Core Ideas in Conservation and Sustainability

Understanding Ecosystems

- Plants and animals can change their environment.
- Things that people do to live comfortably can affect the world around them. People can make choices that reduce their effect on the land, water, air, and other living things.

Source: NGSS Lead States. 2013. *Next Generation Science Standards: For States, by States.* Washington, DC: National Academies Press.

Vignette for Understanding: Recycling Center Explorations

Lisa's class is taking a walking field trip to a recycling center a few blocks from school. Each student is walking alongside a buddy. Each pair of children has a small digital camera that their teachers have given them so that they can document the field trip. Lisa's partner, Jonathon, has the camera first and is taking pictures as the group walks to the recycling center. Once inside, the children switch roles and Lisa takes control of the camera. She first takes several photographs of the large bags of cans that are piled on tables near the entrance to the center. Lisa and Jonathon listen carefully to a man from the recycling center as he talks about how the cans are sorted and cleaned so that they can go on to another center that heats the cans so that they will melt and can be used to make other things.

As the class heads outside, Lisa passes the camera back to Jonathon, and he quickly snaps a picture of a forklift that is parked outside the center. He excitedly tells Lisa that the equipment probably lifts all of the heavy bags of cans inside. The group walks back to the classroom to share their field trip photos with each other. Lisa and Jonathon continue talking about all the cans they saw waiting to be recycled.

Reflection

Field trips can be meaningful ways to extend the knowledge children are building in the classroom. Field trips, whether in-person or virtually on the internet, allow children to connect their understanding to the world outside of the classroom and to build new knowledge about places and events in their communities. This field trip showed Lisa, Jonathon, and the other students the next step in the recycling loop that follows their collections of objects for recycling. Even if you don't have the ability to take your students on a similar trip, you can help children make connections to the steps in the recycling loop or conservation processes that take place outside their homes and classrooms. A great online resource that you can use in your classroom comes from the National Institute of Environmental Health Sciences: https://kids.niehs.nih.gov/topics/reduce/index.htm

Conservation and Sustainability across the Curriculum: Planning Tips

Exploring conservation and recycling topics with your students will provide natural opportunities for you to connect them to other people who work in the school. Children can speak with the school's kitchen and custodial staff to find out how the school works to reduce and recycle every day. These conversations can be connected to early literacy experiences where children are encouraged to write and draw their understanding. As with all science content, science journals will provide an opportunity to share their thinking through their writing and drawing. Photographs from children's direct experiences can be included in the science journals as another method to share their experiences and support their visual communication.

Lesson Ideas

Drawing with Friends: Paper Reuse

Topic:

Scrap paper can be reused in new ways to avoid ending up as trash.

Objective:

The children will learn about reusing paper in collaboration with their classmates.

Materials:

2 bins

Scrap paper

Markers, crayons, colored pencils

Creativity Skills:

Exploration

Visualization

Documentation

Communication/collaboration

Opportunities for unique problem solving

Overview:

Introduce this activity to the whole class. Children will learn about reusing paper and will sort with bins.

1. Activity Steps: Begin by discussing what the children typically do with their scrap paper and unfinished drawings. Let them know that they will now have two new places in the classroom where they can collect these things to reuse them instead of throwing them away.

2. Invite the children to share their understanding of reuse and what it means.

3. Introduce the children to the bin for collecting scrap paper, and talk about how the bin is only for paper that is large enough for someone else to reuse. Talk about how paper scraps must be clean and dry before going into the bin. Show them some example pieces.

4. Introduce the drawings bin. Talk with the children about how they can use this bin to put their incomplete drawings and artwork for other children to take out and reuse.

5. Place bins in an area of the classroom that is accessible to the children, and encourage them to use the paper regularly.

Documentation:

Selecting paper for reuse and reusing paper will serve as a source of understanding about the children's thinking.

Extension Lesson:

Create a space in the classroom to feature children's artwork made from reused paper. Be sure to invite children to create a brief artist's statement to post alongside their artwork so that others will understand how they incorporated the reused paper/art into their artwork.

Reuse Mystery Bag

Topic:

Reusing items can keep them useful and out of landfills, which benefits the land, water, air, and other living things.

Objective:

The children will participate in brainstorming new ways to reuse familiar objects.

Materials:

A large paper or plastic bag

A collection of everyday objects
(at least 1 item per pair of children)

Gift-wrapping paper

Paper lunch bags

Broken toy

Plastic jar

Small cardboard box

Plastic bottle

> **Creativity Skills:**
>
> Exploration
>
> Visualization
>
> Communication/ collaboration
>
> Opportunities for unique problem solving

Overview:

Children will work in pairs to come up with an idea on how to reuse an item from a grab bag.

Activity Steps:

1. Talk with the children as a whole group and let them know that they will be participating in a challenge. Invite them to share what they know about reuse. Separate students into pairs, and invite each pair to come up and pull an item from the grab bag. Once every pair

has an item, encourage them to think about ways they could reuse the item in the classroom. It will be helpful for the children to walk around the classroom as they think about ways to reuse an item. You may want to suggest some options and ideas for reuse.

2. You can encourage them to decorate their object in your art center if desired.

3. Bring the class back together and allow each pair to share their reuse idea.

4. Place the items in the areas of the classroom the children have suggested so that they can be reused.

Documentation:

Informal observations of the children's ideas for reuse, as well as their chosen solution, will illuminate their understanding of reuse.

Extension Lesson:

This lesson can be repeated many times so that the children gain experience in developing reuse ideas and solutions.

Reuse Jetpacks

Topic:

People can reuse materials to reduce their effect on natural resources and other living things.

Objective:

Children will participate in the design and creation of dramatic play items from materials that would otherwise be thrown away.

Creativity Skills:

Exploration

Visualization

Strategic planning

Communication/collaboration

Materials:

Sketch paper and pencils

Pictures of jetpacks or other space equipment

Items for reuse:

　　Soda cans

　　Soda/plastic bottles

　　Scrap paper

　　Scrap string, twine, or cord

　　Scrap cardboard pieces of varying sizes

　　Hot glue gun (adult use)

　　Heavy tape

　　Paint (optional)

Overview:

Children will work in pairs or small groups to create jetpacks by reusing materials in unique ways.

Activity Steps:

1. Invite children to look over the pictures of space equipment, and let them know that they will be working in teams to create jetpacks from items that could have been trash.

2. Encourage the children to sketch plans for their jetpacks. Work with each group or pair to develop a plan. Encourage them to think about the types of materials they will be able to use and how those items will be put together.

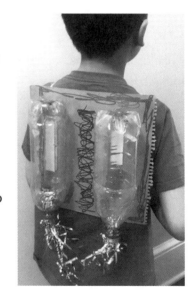

3. Once the groups or pairs have a workable plan, invite them to gather their materials and join you for the creation of the jetpack. As you are working to put their designs together using hot glue, encourage them to refer to their sketches and think about the placement of items.

4. Once the jetpacks are put together and dry, the children can paint or decorate them. Remind them to be careful because the jetpacks are delicate and need to be treated with care.

Documentation:

The children's sketches and final products are one form of documentation. You can also make informal observations and take anecdotal notes about their conversations and ability to communicate and collaborate with their peers.

Extension Lesson:

This lesson can be extended by creating other objects from reusable items, such as found-object artworks.

Biodegradable Flower Planters

Topic:

People can reuse materials to reduce waste, which benefits land, water, air, and other living things.

Objective:

The children will participate in creating biodegradable planters and planting seeds in the planters.

Materials:

Paper-towel tubes cut in half (1 per child)

Potting soil

Flower seeds

Scissors (adult and child)

> **Creativity Skills:**
>
> Exploration
>
> Solution finding
>
> Visualization

Overview:

Students will create a planter made from reused materials. Prior to beginning the lesson, draw 4 lines spaced evenly from one edge of the tube to about 1/3 of the way up. Do this for every tube.

Activity Steps:

1. Talk with the children about how items can be reused, and let them know they will be reusing something that is in most houses and schools: paper-towel tubes.

2. Pass out the tubes and allow the children to look at the lines you've made.

3. Let the children know that they will be cutting along the lines to create a little pot to fill with soil and plant a seed. Once their flower grows, they will be able to put their flower box directly into the ground because the paper tube will disintegrate in the ground.

4. Assist the children in cutting on their lines as needed.

5. Once the tubes have been cut, fold in the cut pieces to form a bottom just like you would close a box.

6. Fill the tubes about 3/4 full with potting soil, and pack it down gently.

7. Have the children add a few seeds and cover lightly with soil.

8. Place the planted pots in a tray, and give them a good watering. You want to completely soak the paper roll and keep it wet the whole time you are growing.

9. Place in a well-lit area that children can easily view in the classroom.

Documentation:

Your informal observations and interactions with the children will serve as documentation of the experience during this lesson.

Extension Lesson:

Extend the lesson by creating other reuse projects, such as recycled paper making.

Virtual Field Trip: Conservation and Sustainability

Topic:

Use digital media to explore how people can make choices that reduce their effect on the land, water, air, and other living things

Objective:

Children will explore child-friendly, Earth-friendly websites.

Materials:

iPad or tablet for the children's use (must have internet access)

Websites for exploration include the following:

- Earth's Kids:
  http://www.earthskids.com/ek_environment.aspx

- Kids for a Clean Environment (group started by a nine-year-old girl):
  http://www.kidsface.org/pages/toc.html

- PBS's Zoom into Action (for kids by kids):
  http://pbskids.org/zoom/activities/action/way04.html

- Environmental Education for Kids (EEK):
  http://www.dnr.state.wi.us/org/caer/ce/eek/earth/air/global.html

Overview:

This lesson works best initially as a whole-group experience so that the children learn how to look at a website and click through a virtual exploration. As they gain experience, you can locate a particular virtual site for them to explore and ask children to work in pairs. Be sure to explore the website you are going to have available for student use ahead of time so that you can be certain the content is appropriate.

Activity Steps:

1. Talk to the children about how many organizations have websites where you can play fun games and watch videos to learn more about reusing, recycling, and reducing the use of items.

2. Choose a website with your students and take the time to explore it.

3. Pause and ask questions so you can gauge their understanding and interest. What do you notice about this picture? Do you think this game will be fun?

4. Once children move to exploring in pairs, periodically check in with them about their virtual experiences. It is a good idea to ask the children to report back to the whole class about what they experienced on their virtual trip and even share an image or two that they found interesting.

Documentation:

Take anecdotal notes about the children's ability to explore, problem solve, and collaborate during their website explorations.

Extension Lesson:

This lesson works best when you combine the content of the websites with other lessons and experiences in your classroom. Be sure to include children's books that reinforce the ideas and content of the children's website explorations in the reading area or during whole-group times.

Children's Books

Child, Lauren, and Bridget Hunt. 2009. *Charlie and Lola: We Are Extremely Very Good Recyclers.* New York: Dial Books.

As part of the Charlie and Lola series, this book takes a close look at how children can recycle and take care of the Earth. It also features lots of recycling tips to get your students involved.

Johnson, Jen Cullerton. 2010. *Seeds of Change: Wangari's Gift to the World.* New York: Lee & Low Books.

Seeds of Change is a biography of the Nobel Peace Prize–winner Wangari Maathai. The story, which focuses on her education, is accessible to children. Your students will learn how Wangari's love of the sciences inspired her life's work to preserve the environment in Kenya.

Trice, Linda. 2016. *Kenya's Art*. Watertown, MA: Charlesbridge.

Kenya's Art shares the story of a young girl inspired to make art from recycled objects after a trip to a recycling exhibit at a museum. The book shares ideas for children to make use of broken toys or other found objects.

5
Earth Science: Earth and Space Systems

Science Experiences that Grow Knowledge of Earth and Space Systems

Young children are excited to learn more about the world around them and the sky above them. Objects in the sky are interesting to children; the clouds and sun during the day and the moon and stars at night are frequently included in children's artwork. During the early childhood years, children should have numerous opportunities to explore the natural world. The Earth's natural resources are easily explored by children in the contexts of their everyday environments—the dirt on the playground, the sand at the beach, the rocks in the school garden, the plants in the garden—are all sources of natural inquiry for children.

Understanding Earth and Space Systems

Earth is made up of land, air, and water, and these resources are used in everyday life. Earth's natural resources—light, air, water, plants, animals, soil, stone, minerals, and fossil fuels—can be conserved for future generations. All the plants and animals that live on Earth are affected by the changes in weather that occur daily and during each season. Weather can be predicted by scientists known as *climate scientists* and *meteorologists*.

Their work helps prepare us for changes in weather that can affect daily life.

The night sky allows children to see the moon and stars, while the day sky allows children to see the sun, clouds, and the moon. The study of the Earth and space includes understanding the mechanics of day and night and the phases of the moon. There are four major phases of the moon: new moon (occurs when the moon is positioned between the Earth and sun, allowing us to see only a slender crescent); full moon (the moon is on the opposite side of the Earth so the entire sunlit part of the moon is facing us); and the first and third quarter moons (often referred to as half-moons, these occur when the moon is at a 90-degree angle to the Earth and sun so half of the moon is illuminated and the other half is shadowed).

The Cycle of Inquiry: Earth and Space Systems

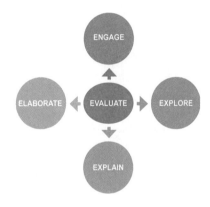

Engage

Questions to Engage: Earth and Space Systems

Ask questions such as the following to engage young learners' interests:

- Can you see the moon during the day?
- Why is your shadow so tall right now?
- How are these rocks the same? How are they different?
- Can you describe this rock with one word?

Explore

Engaged Exploration: Earth and Space Systems

To assist children as they work to explore the world around them, provide a space in the classroom that invites children to do the following:

- Sort, compare, and classify natural Earth objects by their physical properties (color, shape, texture, size, and weight)
- Explore the changes that can be observed in the day and night sky
- Explore the changes in weather over time

Explain

Explanation Opportunities: Earth and Space Systems

Use the following suggestions to help children understand and explore class lessons on a deeper level:

- Display properties charts of Earth materials
- Plan whole-group debriefings that discuss observed changes in weather
- Encourage students to use science journals for drawings, photographs, and student comments of long-term studies of the night sky and the changing environment
- Develop whole-class exploration charts including weather charts, the sun and shadows throughout the day, and the phases of the moon

Elaborate

Science Center Elaborations: Earth and Space Systems

Keep your students engaged with the following ideas to continue learning in the classroom:

- Develop classifying opportunities with a variety of Earth materials to explore the size, shape, color, texture, or hardness properties

- Use flashlights to explore shadow

- Display images of the Earth and sky for children to use in their research

Evaluate

Documenting Informal Evaluation: Earth and Space Systems

Help your students evaluate their learning from the explorations by using the suggested ideas:

- Plan whole-class review sessions

- Have a science night for families and friends

- Display documentation panels of major projects

- Put together student portfolios

- Involve students in the planning of next steps based on their thoughts and ideas

Core Ideas in Earth and Space Systems

Understanding Earth and Space Systems

Patterns of the motion of the sun, moon, and stars in the sky can be observed, described, and predicted.

Understanding Seasonal Patterns

Seasonal patterns of sunrise and sunset can be observed, described, and predicted.

Earth Science: Earth and Space Systems

Understanding Weather and Climate

Weather is the combination of sunlight, wind, snow or rain, and temperature in a particular region at a particular time. People measure these conditions to describe and record the weather and to notice patterns over time.

Source: NGSS Lead States. 2013. *Next Generation Science Standards: For States, by States.* Washington, DC: National Academies Press.

Vignette for Understanding: My Shadow

Julie's class has been spending a lot of time learning about the sun, stars, and the moon. They are working with a partner on the playground to create a chalk outline of each child's shadow. Julie is standing very still in an exaggerated pose while her friend Kailani traces Julie's shadow on the concrete section of the school's playground. Kailani announces that she's finished tracing Julie's shadow, so Julie turns around to see her shadow. Both girls laugh at the result and Kailani says, "Okay, my turn. I'm going to stand next to your shadow so our shadows can play." Kailani and Julie carefully line up Kailani's shadow with Julie's tracing. Once they are satisfied with the placement, Julie takes a piece of chalk and begins to trace. The girls take the time to color in their shadows and walk around the playground to see the outlines their classmates have made.

Reflection

The girls were able to explore their shadows as an extension to previous lessons and experiences where they learned about the sun and its movement during the day. They were able to work together and create a plan for their drawings without needing assistance from their teachers. It's important to find a balance in your science experiences between teacher-assisted experiences and children's independent experiences. Often, science experiences require more teacher assistance because of the materials used or the complexity of the steps. Providing opportunities in which children can work together during science experiences without your direct involvement will help support their content-knowledge growth and their confidence in their abilities.

Earth and Space Systems across the Curriculum: Planning Tips

Experiences with Earth and space systems will likely take place over longer periods of time, as a key idea of the content involves an understanding of change over time. Opportunities to integrate math, social studies, the arts, and early literacy skills will be plentiful as children create project journals that explore seasonal change, weather, and the night sky. Encourage your students to use their journals to reflect on what they've observed and learned over time. Long-term work in the classroom can be fun to share with parents and families; create opportunities to share children's project documentation over the course of the projects and not just at the project's completion. Getting families involved along the way will help support children's developing knowledge.

Long-Term Project: At-Home Moon Journals

Topic:
Patterns of the motions of the sun, moon, and stars in the sky can be observed, described, and predicted.

Objective:
Children will observe and document the changes of the moon over the course of a month.

Creativity Skills:

Exploration

Communication/collaboration

Documentation

Materials:

Science journal (enough pages for children to draw nightly for a month)

Pencils

Overview:

This activity will be done at home, so you will need to create a simple science journal (a space to note the date and a large space for drawing the moon) that can be sent home. You will also need to speak with parents or send a note home about the goals for the project. Plan to do this project in the late fall when Daylight Savings Time ends, as the moon will be visible before children will need to be in bed.

Activity Steps:

1. Ask parents to find a few minutes every night to observe the moon with their child and encourage the child to draw his or her observation in the journal. Parents can even write down their child's thoughts about how the moon looks or maybe why they couldn't see the moon that evening (cloudy sky). Assure parents that it is okay if they miss some nightly recordings—the idea is to follow the moon for a month to explore the changes over time.

2. Once the month has ended, invite the children to bring their journals back to school to share. Have children show their drawings on the same dates to compare with their classmates' observations.

3. Display the children's drawings in the classroom.

4. Invite the children to share their observations about the changes that occurred over the course of the month.

Documentation:

The children's journals are used as a source of documentation. Be sure to follow up with parents about their experiences with the project; their reflections will also be a source of information for you.

Extension Lesson:

This lesson can be extended by helping the children label their moon drawings with the phases of the moon. Charts of the moon phases can be hung in the classroom to reinforce this information.

Long-Term Project: *Seasonal Change*

Topic:
Seasonal patterns can be observed, described, and predicted.

Objective:
Children will observe and document the changes of a tree in the fall.

Materials:
Science journal

Colored pencils, crayons, markers

Digital camera (optional)

Clipboards (optional)

Creativity Skills:
Exploration

Visualization

Communication/collaboration

Documentation

Overview:
This activity can be done as a whole group and will take place over four to six weeks in the fall. You will need to select trees that will change color and are on your school property or nearby. To see a list of trees that change color, visit www.arborday.org/shopping/trees/topfalltrees.cfm

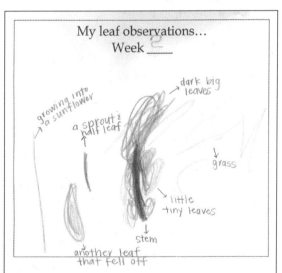

Earth Science: Earth and Space Systems

Activity Steps:

1. In advance, create simple science journals that include spaces for children to write or draw their weekly observations for four to six weeks. You can also create a space for children to make predictions about what they will see the following week.

2. Prior to beginning your observations of the trees, invite the children to talk about what happens in the fall: Can you tell me what happens to leaves on a tree in the fall? Do you know why that happens? What are some of the colors that leaves can change to?

3. Choose one day a week when you can take the children outside to make their observations and record them in their science journals. Allow at least 20 minutes for exploration and recording. It is important for the children to be able to record their observations while they are outside observing the trees and leaves.

4. Invite the children to choose one tree to observe each week. Encourage them to draw what they are observing. You can ask prompting questions such as, What colors are you seeing on the trees' leaves? Do you notice any differences in color this week? Have you looked at your drawing from last week? What is different this week?

5. You can supplement their drawings with descriptive text that the children dictate to you, and you can include weekly photographs of the tree in their journals.

6. At the end of the project, invite the children to share their journals with the class and their families.

Documentation:

The children's journals are documentation, and so are the conversations you have with children each week while they are drawing.

Extension Lesson:

This lesson can be extended by doing a similar project in the spring and tracking the emergence of leaves on the trees.

Long-Term Project: Weather Changes

Topic:

Weather is the combination of sunlight, wind, snow or rain, and temperature in a particular region at a particular time. People measure these conditions to describe and record the weather and to notice patterns over time.

Objective:

Children will observe and document weather changes over the course of a month.

Materials:

Weather journal (enough pages for children to draw daily at school for a month)

Colored pencils, crayons, markers

Outdoor thermometer

Windsock

Rain gauge

Creativity Skills:

Exploration

Communication/collaboration

Documentation

Earth Science: Earth and Space Systems

Overview:

This activity can be done as a whole group and will take place over a month during one season. Select the season in which you think the children will be able to observe a variety of weather conditions. This experience works best if you plan daily observations at the end of the children's outside playtime.

Activity Steps:

1. In advance, create a simple science journal page that includes space for children to write down the temperature, wind conditions, and rain measurement. Include a space for them to draw their observations of the daily sky.

2. Place an outdoor thermometer, windsock, and rain gauge outside near the school classroom. Talk with the children about how these tools are used by weather scientists to track the changes in weather over time.

3. Let the children know that they will be tracking the weather for one month using these tools.

4. Set aside 10 minutes for them to record their observations each day while they are outside and near the weather equipment. On inclement-weather days, children can make their observations while indoors.

5. Encourage the children to make comparisons between the previous day's weather and the current day's observations.

6. Once the project is completed, invite the children to share their weather journals with the other students and their families.

Documentation:

The children's journals are documentation, and so are the conversations you have with children each day while they are drawing and recording the weather facts.

Extension Lesson:

This lesson can be extended by completing a weather journal for each season. The journals can be used to compare weather across seasons to notice the changes that occur in precipitation and temperature.

Children's Books

Cole, Rachel. 2017. *City Moon*. Toronto: Schwartz and Wade.

With colorful, collage-style artwork, *City Moon* tells an engaging story that your students will connect with as the main characters search for the moon on an evening walk.

Minor, Wendell. 2015. *Daylight Starlight Wildlife*. New York: Nancy Paulsen Books.

This book connects life and Earth science by introducing your students to nocturnal (active at night) and diurnal (active during the day) animals. Each creature is introduced through lyrical text and detailed paintings.

Sayre, April Pulley. 2015. *Raindrops Roll*. San Diego, CA: Beach Lane Books.

With bright, colorful photographs from all parts of world, *Raindrops Roll* tells the story of what raindrops do and how rain affects the natural world.

Index

5E inquiry model, 3
   Earth and space systems, 87–89
   growth and change, 55–57
   matter and physical properties, 18–20
   physical and chemical changes, 36–38
   steps, 5–8

A
Allergies, 60–61, 63
Aluminum foil, 25
Animals, 54–55, 68–69
   ecosystems, 57, 74, 86–87
   habitats, 1–3, 69
Art center, 78–79
Art skills. *See* Visual-art skills
Atoms, 35

B
Bags, 44, 78
Baking soda, 41, 49, 51
Balloons, 49
Bins, 27, 76
Bottles, 29–30, 34–35, 37, 49, 78, 80
Bowls, 42, 44, 52, 61
Boxes, 78
Brainstorming, 11–12, 78–79
Building blocks, 23
Buoyancy, 17–18, 26–28
Bybee, Rodger W., 5, 16

C
Cabochons, 23, 30
Cameras, 61, 63, 65, 74–75, 93
Cans, 80
Carbonated cola, 34–35
Card stock, 25
Cardboard, 80
Celery, 62–65
Cellophane, 25
Chemical changes
   properties of, 38
Chemical properties
   defined, 35
Children's books, 15–16
   Ada Twist, Scientist by Andrea Beaty, 32
   All about Matter by Mari Schuh, 33
   Bartholomew and the Oobleck by Dr. Seuss, 52
   Charlie and Lola: We Are Extremely Very Good Recyclers by Lauren Child & Bridget Hunt, 84
   City Moon by Rachel Cole, 97
   Daylight Starlight Wildlife by Wendell Minor, 97
   If I Had a Gryphon by Vikki VanSickle, 3
   If My Mom Were a Platypus: Mammas Babies and Their Mothers by Dia Michels, 70
   Kenya's Art by Linda Trice, 85
   Little Seeds by Charles Ghigna, 69
   Miss Maple's Seeds by Eliza Wheeler, 70
   Pancakes, Pancakes! by Eric Carle, 52
   Pie in the Sky by Lois Ehlert, 69
   Pop! A Book about Bubbles by Kimberly Brubaker Bradley, 51
   Raindrops Roll by April Pulley Sayre, 97
   Seeds of Change: Wangari's Gift to the World by Jen Cullerton Johnson, 84
   What's the Matter in Mr. Whiskers' Room? by Michael Elsohn Ross, 33
Classifying skills, 19–20, 22, 28–30, 37–38, 43–46, 60, 73, 88–89
Classroom environment, 13, 15
   art center, 78–79
   sand table, 22, 43
   science center, 13–15, 22, 60–62, 65, 88–89
   water table, 22, 27, 43
Climate science, 86–87, 90, 95–97
Clipboards, 93
Collaboration skills, 4, 15, 21–32, 39–51, 60–69, 76–84, 91–97
Color wheels, 65
Colors, 17, 19, 22–24, 28–30, 35–36, 44, 46–48, 51, 62–65, 88–89, 93–95
   mixing, 21–22
Communication skills, 4, 10, 23–26, 28–30, 30–32, 41–51, 60–69, 76–81, 91–97
Comparing/contrasting, 18, 87–88, 96
Conservation and sustainability, 16, 71–85
   children's books, 84–85
   core ideas, 74–75
   cycle of inquiry, 72–74
   lesson ideas, 76–84
Containers, 25
Content coverage, 12–13
Cooking activities, 41–43, 52–53
Cooking oil, 25, 28, 30–31, 37, 41
Cords, 80
Core ideas
   conservation and sustainability, 74–75
   Earth and space systems, 89–91
   growth and change, 57–59
   matter and physical properties, 20–21
   physical and chemical changes, 38–40
Corn syrup, 30
Cornstarch, 28, 44, 52
Counting, 29, 31
Craft sticks, 23
Craft, Anna, 12, 16
Crayons, 23, 65, 76, 93, 95
Creative-thinking skills, 1–3, 11–14, 17
Critical-thinking skills, 11–12
Crystals, 39–40
Cups, 65
Curiosity, 3
Cutting boards, 61
Cycle of inquiry
   conservation and sustainability, 72–74
   Earth and space systems, 87–89
   growth and change, 55–57
   matter and physical properties, 18–20
   physical and chemical changes, 36–38

D
Deep thinking, 9–10, 15, 60
Descriptive language, 10, 18, 20, 23–25, 28–30, 35, 38, 43–46, 87, 89, 91–94
Digital media, 83–84
Documentation, 4, 14, 20, 24, 26–30, 32, 38, 41–43, 45, 47, 49–51, 60, 62–68, 73–77, 79, 81–82, 84, 89, 91–97
Dramatic play, 79–81
Driftwood, 27

E
Earth and space systems, 16, 86–97
   children's books, 97
   core ideas, 89–91

cycle of inquiry, 87–89
lesson ideas, 91–97
Earth materials, 15, 24, 27, 86–89
Earth sciences, 3, 11, 16
conservation and sustainability, 71–85
Earth and space systems, 86–97
Ecosystems, 57–59, 86–87
understanding, 74
Engineering experiences, 3, 14
Epsom salts, 39–40
Evaluation, 7–8
Exploration/experimentation, 3–4, 7, 11–12, 15, 17
conservation and sustainability, 76–84
Earth and space systems, 91–97
growth and change, 60–69
matter and physical properties, 21–32
physical and chemical changes, 40–51

F

Fair tests, 31–32
Family participation, 14, 20, 38, 40, 57, 74, 89, 91–93, 96
Field trips, 74–75
virtual, 83–84
Flashlights, 89
Flour, 41–42
Following directions, 43
Food coloring, 28, 30, 39–42, 44, 46, 51–52, 63
Formative assessment, 15
Funnels, 49

G

Gardening activities, 55–56, 65–68, 81–82
Gases, 35–37, 41–43, 49–51
defined, 35
Gift wrap, 78
Glasses, 46
Glitter, 37
Glue, 30
Gravity, 17
Growth and change, 16, 54–70
children's books, 69–70
core ideas, 57–59
cycle of inquiry, 55–57
lesson ideas, 60–69
Guided inquiry, 5–8

H

Habitats, 1–3, 69
Hardness, 17, 19, 35, 89
Honey, 31
Hot glue guns, 80

I

Immersion, 13
Innovation, 13
Inquiry-based learning, 1–3, 17
promoting, 12–13
iPads, 83–84

J

Jars, 30, 63, 78
Journaling. See Science journals
Juice, 48

K

Knives, 61

L

Leaves, 27

Lesson ideas
conservation and sustainability, 76–84
Earth and space systems, 91–97
growth and change, 60–69
matter and physical properties, 21–32
physical and chemical changes, 40–51
Life sciences, 3, 11, 16
growth and change, 54–70
Liquid soap, 25
Liquids, 24–26, 28–32, 37, 41–49
defined, 35
non-Newtonian, 53
Literacy skills, 3, 76, 91, 93

M

Magnetism, 17, 19, 28
Magnets, 28
Magnifying glasses, 15, 61, 65
Manipulation skills, 3, 41–43
Marbles, 30
Markers, 63, 65, 76, 93, 95
Mass, 17
Matching skills, 68–69
Math counters, 23
Mathematics, 3, 22, 91
Matter and physical properties, 16, 17–33, 38
children's books, 32–33
core ideas, 20–21
cycle of inquiry, 18–20
definitions, 17, 35
lesson ideas, 21–32
Matthew, Alice, 16
McConnon, Linda, 16
Measurement skills, 39–46, 49–51, 56, 60, 65–68
temperature, 95–97
Measuring cups, 44
Measuring spoons, 44
Medicine droppers, 25, 50–51
Melting points, 17–18, 21–22, 35–36
Mentos candy, 34–35
Meteorology, 86–87, 95–97
Microwave ovens, 44
Milk, 48
Minds-on learning, 3–4, 7, 34–35
Misconceptions, 4, 6, 8–10, 34–35
Molasses, 31
Molecules, 35, 46–49
Moon phases, 87–93

N

National Academies Press, 12
Next Generation Science Standards, 3, 12, 20, 38, 57, 74, 90

O

Observation skills, 1–4, 10, 17, 23–24, 30–32, 39–40, 49-51, 56–59, 62–69, 73, 76–82, 86, 88–95
Oobleck, 52–53
Open-ended tasks, 11
Outdoor activities, 14, 19, 26–28, 95–97

P

Paint, 50, 80
Paper, 25, 31, 76, 79–80
making, 82
parchment, 25
Paper sacks, 23
Paper towel tubes, 81
Parents. *See* Family participation

Peer-to-peer relationships, 14–15, 21–22, 60–62, 68–69, 78–84, 90–91
Pencils, 27, 31, 61, 63, 65, 76, 79, 92–93, 95
Photos, 19, 36, 62, 68, 79
Photosynthesis, 54–55
Physical and chemical changes, 16, 34–53
    children's books, 51–52
    core ideas, 38–40
    cycle of inquiry, 36–38
    definitions, 35
    lesson ideas, 40–51
    understanding, 34–36
Physical characteristics, 1–3
Physical properties, 88–89
    defined, 35
Physical sciences, 3, 11, 16
    matter and physical properties, 17–33
    physical and chemical changes, 34–53
Pinecones, 27
Pipettes, 25, 50–51
Plants, 54–57, 60–62, 65–68, 81–82, 86–87
    ecosystems, 74
    exploring roots, 58–59, 62–65
    seasonal changes, 93–95
Plastic wrap, 25
Playdough recipes, 41–43
Playful learning, 4–5, 12, 15
Possibility thinking, 12
Powdered drink mix, 42
Predicting skills, 17, 25–26, 30–32, 35, 39–40, 42, 47, 55, 58–62, 64, 89, 91–93, 95–97
Prior knowledge, 6–7, 64, 94
Problem-solving skills, 4, 11–12, 14, 26–28, 41–51, 76–79, 84
Project-based learning, 4, 14–15, 39–40, 62–68, 79–81, 91, 93–97
Properties charts, 19, 88

Q
Questioning, 1–3, 5, 7, 10, 12–13, 15, 18, 24, 35, 42, 55, 58–59, 61, 72, 84, 87

R
Rain gauges, 95
Recipes
    fizzy playdough, 42–43
    no-cook playdough, 42
    oobleck, 52–53
    playdough, 41–42
    powdered drink mix playdough, 42–43
Recommended practices, 12–13
Recycling, 72–76, 83–84
    defined, 71–72
Reducing, 72–74, 76, 83–84
    defined, 71
Reusing, 72–74, 76–82
    defined, 71
Risk taking, 12–13, 15
Rocks, 27
Rulers, 15

S
Salt, 41–42
Sand, 27, 37
Sand table, 22, 43
Scaffolding, 3
Scales, 15
Science center, 13–15, 22, 60–62, 65, 88–89
Science journals, 15, 19, 27–28, 36, 56, 61–63, 65, 76, 88, 91–93, 95

Science language, 26–28
Scissors, 25, 63, 81
Sea glass, 27
Seashells, 27
Seasonal patterns, 89, 91, 93–95, 97
Seeds, 27, 60–62, 65–68, 81
Self-determination, 11–13
Shadows, 9–10, 87, 89–91
Shape, 17, 19, 22–24, 35, 88–89
Size, 17, 19, 22–24, 27, 35, 88–89
Social skills, 11
Social studies, 3, 91
Soil, 65, 81
Solids, 35–37, 41–46, 49
    defined, 35
Solution finding, 26–30, 46–48, 68–69, 81–82
Sorting skills, 18, 22–24, 60–62, 73, 76–77, 88
Spoons, 44, 52, 61
Stones, 23
Strategic planning skills, 41–43, 68–69, 79–81
String, 80
Student portfolios, 20, 38, 57, 74, 90

T
Tablets, 83–84
Taking turns, 23–24, 90–91
Tape, 80
Teachers
    creating environments, 11–13
    role of, 3–5, 15
Technology, 3, 14, 83–84
Texture, 17–19, 22–24, 35, 88–89
Thermometers, 95
Tiles, 23
Timers, 46
Tools, 58–59, 65
Towels, 27, 61
Toys, 27, 78
Trees, 93–95
Twigs, 27
Twine, 80

V
Vases, 63
Vinegar, 42–43, 50–51
Visual-art skills, 1–3, 23–24, 30–32, 39–40, 62–69, 73, 76–82, 86, 91, 93–95

W
Water, 22, 25, 27–28, 30, 37, 39–44, 46, 52, 61, 63
Water play, 18–19
Water table, 22, 27, 43
Wax paper, 25
Weather, 86–88, 90–91, 95–97
    charts, 88
    misconceptions, 9–10
Websites
    Arbor Day, 93
    Earth's Kids, 83
    Environmental Education for Kids, 83
    Kids for a Clean Environment, 83
    National Institute of Environmental Health Sciences, 75
    PBS's Zoom into Action, 83
Weight, 17–18, 23–24, 27, 88
Windsocks, 95

Z
Ziplock bags, 28
Zoomorphic animals, 1–3